Stillwater Flies BOOK 1

TAFF PRICE'S

Stillwater Flies BOOK 1

A Modern Account of Natural History, Flydressing and Fishing Technique

ERNEST BENN LIMITED
London and Tonbridge

Published by Ernest Benn Limited
25 New Street Square, London EC4A 3JA
& Sovereign Way, Tonbridge, Kent, TN9 1RW

First Published 1979

Printed in Great Britain by
Ebenezer Baylis & Son Ltd,
The Trinity Press, Worcester & London

ISBN 0 510-22525-X

Contents of Book 2

Contents of Book 3

Other Benn Books on Flydressing

JOHN GODDARD
The Superflies of Still Water

T. DONALD OVERFIELD
Fifty Favourite Nymphs

TOM STEWART
Two Hundred Popular Flies and how to tie them

RICHARD WALKER
Fly Dressing Innovations

Stillwater Flies BOOK 1

The Midges

ORDER: Diptera
Family: Chironomidae

The delicate and oft times written about Ephemeroptera (the Mayflies) may well hold their place as the royal flies of the chalk stream, but on still water the more plebeian Diptera are the most important source of food for the trout. If the chalk-stream is the kingdom where the Mayfly reigns supreme, then the lakes and reservoirs are republics where the Chironomid midge, in its many forms, governs.

We have in the British Isles about 400 species of the family Chironomidae. The one that is of greatest interest to the angler is *Chironomus plumosus*, although all the midges can pose an identification problem. However, providing we match size and, to a degree, colour, then that is usually sufficient to meet our humble needs, for we are at the water to fish and not always to try and count the number of hairs on an insect's left legs.

The adult midge can vary in size from about 5mm to 13mm depending on the species. It is the simplest thing in the world to ascertain the size of the local variety by using our God-given eyes, by noting what insect is coming off the water, or even by looking at what insects are resting on the bankside vegetation. It is surprising what we can learn from just a few moments' observation of nettles, bushes and other waterside plants.

Life Cycle

Most species lay their eggs in a jelly-like mass which quickly sinks to the bottom of the lake or reservoir. They soon hatch out into worm-like larvae. The larvae of the midge varies in colour with each individual species; one of the best known is the bloodworm, aptly named for it carries in its system haemoglobin, giving it its blood red coloration. It needs haemoglobin for much the same reason as humans need it, to carry oxygen around its system. The bloodworm's habitat is usually in the deepest part of the water, down in the mud where oxygen levels are practically non-existent. Other larvae are coloured green, some are cream, and others are of a yellow coloration. These larvae construct for themselves 'U' shaped tunnels, and live on planktonic matter and algae.

The next stage in the metamorphosis is the pupa, and from

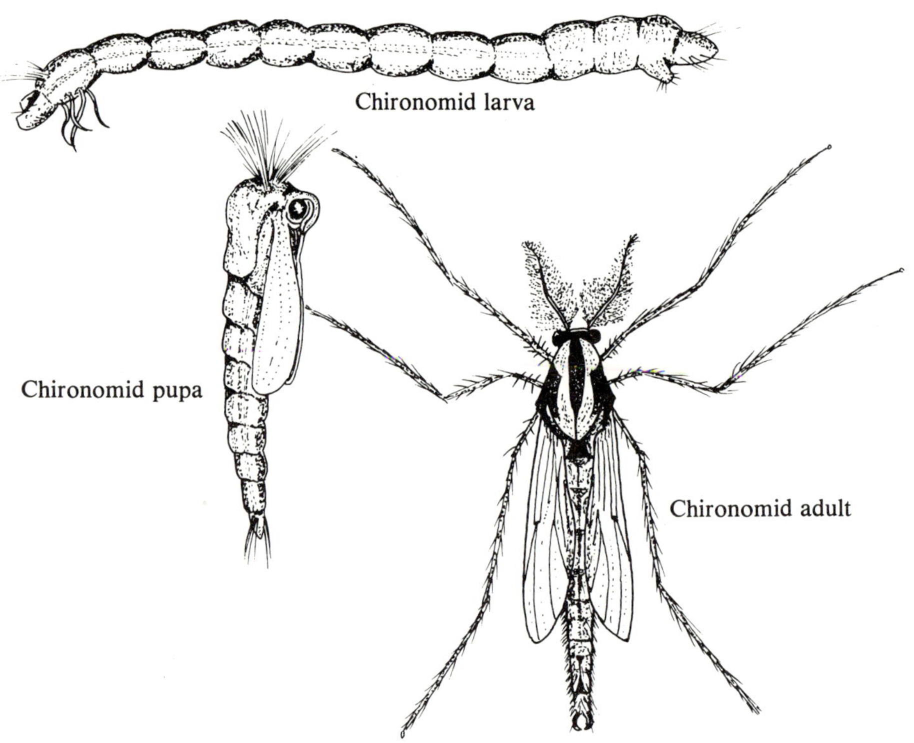

an angling point of view this is the most effective stage to imitate, for many of the pupae are mopped up by the trout, either on their journey to the surface or when they hang hook-like in the surface film, struggling to break through the surface tension of the water. When they finally break through to the surface the pupal skin splits and the adult fly emerges. As it struggles out of its shuck, in many instances the wings take on a distinctly orange hue, which fades within a minute or so.

The sole purpose of the adult midge's life is to reproduce—a short life, but a happy one! The males can often be seen dancing in clouds near to the water. This undulating dance was the origin of the old name for these flies, Dancing Gnats. This smoky cloud of males is joined by individual females which have been resting up on nearby vegetation; the mating usually takes place an hour or so after hatching, though on one occasion I witnessed the pairing actually on the water as soon as the females hatched out.

As a point of recognition, the male midges have long plumed antennae. The females' antennae are short and unplumed.

Surridge Wormbody
This is a material fairly new to flydressers, created by Graham Surridge. It is comprised of a tight narrow spiral of nylon, an excellent medium for tying quick, effective larval and pupal imitations; the red version makes a very good bloodworm larva.

Larval Imitations

Marabou Bloodworm
Hook L/S 12-14
Silk Crimson
Tail A tuft of red marabou (this is in effect a body extension rather than a tail)
Body Segmented red floss silk
Rib Fluorescent red silk
Head (Optional) Peacock herl

Tying notes

1. Tie in a bunch of marabou fibres for the tail, followed by the fluorescent rib silk and then the red body floss.
2. Wind the body floss in undulating turns, leaving room for the head if desired.
3. Rib with the fluorescent red silk, and tie in peacock herls for the head.
4. Form the head, tie off the peacock herls, and whip finish.

Marabou Green Larva
As for bloodworm, but use olive green materials

Marabou Yellow Larva
As above, but yellow

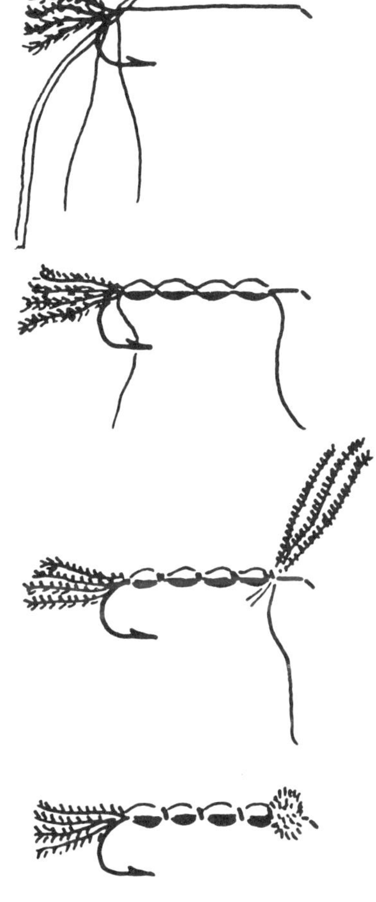

Tying a Marabou Bloodworm

Other Larval Patterns

Red or Green Larvae
(J. Goddard)
Hook L/S 8-12
Silk Brown
Tail Curled red ibis substitute or green goose
Rib Narrow flat tinsel
Head Buff condor herl

Rubber Band Fly (A. Cove)
Hook 12-14
Body A curved section of rubber band (detached)

Richard Walker further developed this by using crimson swan herl for the body ribbed with clear monofilament nylon.

Red Bloodworm
(J. T. Lane)
Hook 12-10
Silk Blood red
Body (Detached) Blood red floss tied on top of hook and trailing for about $\frac{3}{4}$in. in length

Pupal Imitations

As stated earlier, the pupa is the most killing of the midge's forms, from an angling point of view. All anglers and fly-dressers have their own favourite patterns designed to imitate this stage in the midge's life, and quite honestly there is not much to choose between any of them—they are all very similar in shape if not in construction. The only difference between mine and the others is the fact that I have given the fly a body extension of marabou fibres in order to simulate the twitch and wriggle of the natural creature as it struggles in the surface film.

Marabou Midge Pupa
(Price)
Hook 10-14
Tail Marabou feather (black, orange, olive, etc.)
Body Floss silk
Rib Fine oval silver tinsel
Thorax Bronze peacock herl
Breathing Filaments Soft white feather fibres

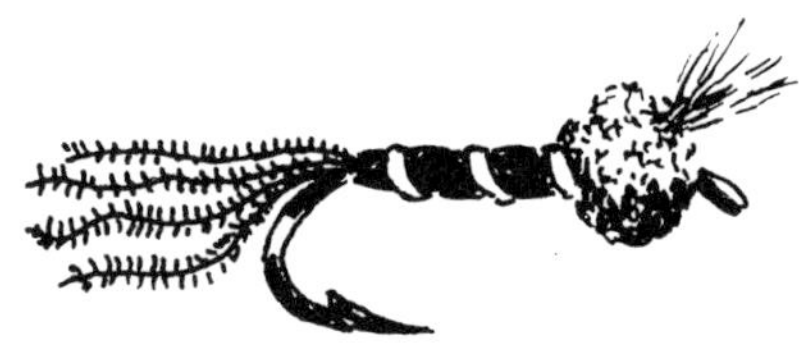

Marabou Pupa

Other Pupal Patterns

Footballer (G. Bucknall)
Hook 10-18
Body Black and white horse hair wound together
Thorax Mole's fur
Head Peacock herl

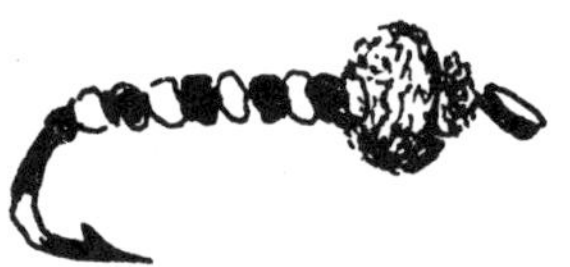

Bucknall's Footballer

Hatching Midge Pupa
(D. Collyer)
Hook 6-12
Body Floss silk of appropriate colour
Rib Flat gold or silver tinsel
Thorax Clipped deer's hair
This fly utilises clipped deer's hair for its buoyant properties, allowing the fly to hang in the surface film.

Collyer's Hatching Midge Pupa

Hatching Midge Pupa
(J. Goddard)
Hook Straight eyed 10-14
Silk As the desired body colour
Body Floss silk ribbed with silver lurex lapped over with PVC
Tail Short tuft of white feather fibres
Thorax Peacock or buff condor herl
Breathing Filaments White hackle fibres

Goddard's Hatching Midge Pupa

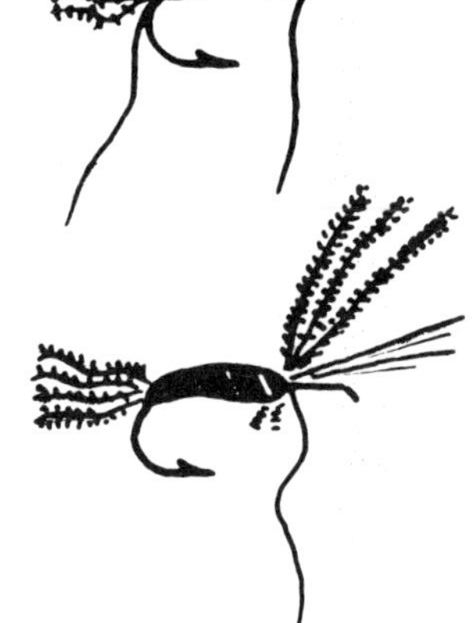

Tying a Marabou Midge Pupa

Tying notes
1. Tie in marabou fibres for the tail, followed by the silver rib tinsel and the body floss.
2. Wind on the body floss, leaving room for the thorax.
3. Rib with the silver tinsel, then tie in the breathing filaments and thorax herls.
4. Wind on peacock herl thorax, and trim off the filaments. Whip finish.

Adult Midge Imitations

So much for the pupa. There should be enough dressings there for the most fastidious to choose from. Now for the adult fly.

I am of the opinion that there are three distinct taking times when the midge is on the water. The first is as it is in the process of hatching; this I have termed the emergent midge. The second is the resting stage, and the third is when the midge flies rapidly over the water with its legs just touching the surface, appearing to whirl around. This stage I term the Skating or Whirling Midge.

Emergent Midge

Hook	Up eyed 10-14
Tail	Small grizzle hackle (this imitates the pupal shuck)
Body	Polypropylene or seal's fur tied thinly
Rib	Fine gold tinsel
Wing	Dyed orange swan, tied down at head and tail over the back of the fly
Thorax	As for body
Hackle	Cock hackle clipped underneath

The colour of this fly can be black, grey or olive

Resting Midge

Hook	Up eyed 10-14
Body	Polypropylene dubbing or seal's fur
Rib	Fine gold tinsel
Thorax	As for body
Wing	Grey hackle fibres flat across the back
Hackle	Cock hackle of appropriate colour

Skating or Whirling Midge

Hook	Up eyed 14
Body	As for other two patterns
Rib	Gold oval tinsel
Hackle	Cock hackle of good quality almost two inches in fibre length

This latter fly is a derivation of the American Skater pattern devised by Edward Hewitt

Fishing the Midge

The midge is in evidence throughout the fishing calendar. There are not many days when it does not hatch; sometimes the hatch is very sparse, at other times they come off the water in vast numbers. As for time of day, again there are no hard and fast rules. I have seen good hatches provoke trout into feeding early in the morning, likewise in the heat of the afternoon in high summer, as well as at dusk.

Larva Fish close to the bottom either with a very long leader and floating line, or with a sink tip or slow sinker. Fish the fly with long slow pulls and with an occasional twitching of the rod to activate the tail. This is particularly effective with the Marabou Bloodworm. After rough weather the larva can be fished at all depths, for many naturals are washed out of their tunnels at such times.

Pupa Fish with a well greased leader, in the surface film, with hardly a retrieve at all, allow the natural water drift (or the wind) to move your fly. Again, if using a marabou version, an occasional uplifting of the rod twitches the tail into life.

Adult Midge Allow the Resting Midge to float with the drift or breeze. The Emergent Midge should be fished in the surface film, jerking the rod from time to time to simulate the struggle of the natural fly as it tries to leave its pupal shuck. The Skating Midge should be quickly skimmed across the water, with occasional pauses.

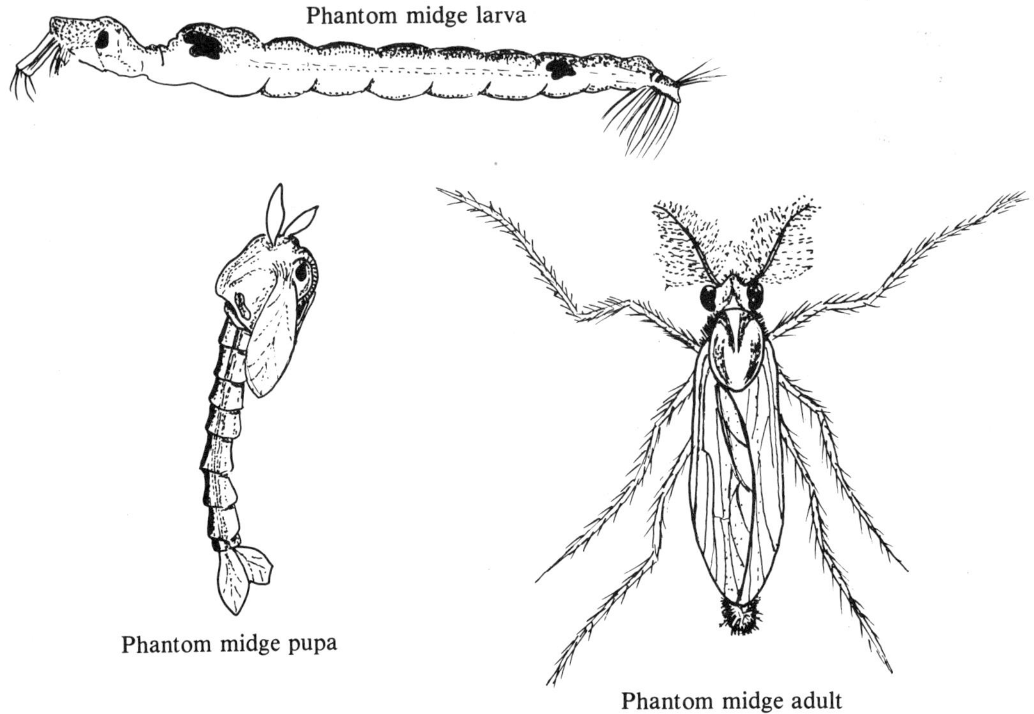

Phantom midge larva

Phantom midge pupa

Phantom midge adult

The Phantom Midge

ORDER: Diptera
Family: Chaoborinae

The Phantom Midges belong to a family called Chaoborinae. At one time they were included with the mosquitoes (Culicidae). Unlike the mosquitoes, they do not bite: I suppose one could place the Phantoms between the biting mosquitoes and the non-biting chironomids.

How important are these ghostly flies to the angler? Probably more important than we imagine, for this creature lives in still water (in its larval and pupal stages) in vast numbers, and so it must figure highly in the diet of the trout.

The Chaoborus midge is called the Phantom Midge (or Phantom Fly) from the fact that its larva is almost completely transparent, and hovers in the water like a ghostly submarine.

Life Cycle

The female midge lays her eggs on the surface of the water. Some entomologists believe the eggs hatch out on the surface, the infant larvae fluttering down to the lake or reservoir bottom. Others are of the opinion that the eggs themselves sink to the bottom, and hatching takes place there.

The larva of the Phantom Midge differs from the chironomids inasmuch that it is free swimming, and again, instead of having breathing spiracles the phantom larva breathes directly through its skin, absorbing oxygen from the water. It feeds on planktonic matter; hovering, it lies in wait for its prey with an uncanny stillness. When it wants to, however, it can put on a fair turn of speed, by a quick contortion of its body. One minute you see it, then you don't, and as if by magic it reappears a few inches away, hovering with the same ghostly intent.

The next stage in the life cycle is the pupa, again a creature that is almost transparent, but not quite as invisible as the larva, for it has a slight watery green colour, which darkens prior to hatching to a pale browny orange. The pupa can hibernate through the winter, becoming active once more when the water temperature increases; then it rises to the surface, the skin splits and the adult fly emerges, to fly off to the shade of the bankside vegetation.

General Identification

The accompanying sketch of all three stages will give the reader some idea how this insect looks.

David Jacques, *The Development of Modern Stillwater Fishing*, A. & C. Black.

Adult A small, mosquito-like fly (size 6-8mm) resembling in many ways the 'buzzer' that most stillwater anglers are acquainted with. It is usually of a pale watery green colour. The males have plumed antennae (in some areas this insect was called the Plumed Gnat). When at rest the wings, which are hairy, are held flat over the back, covering most of the abdomen, and the fly tilts forward.

Larva As already stated this stage (size 10-12mm) is virtually transparent except for a couple of dark spots, one near the head, the other further down towards the end of the abdomen. These are thought to be stabilisers, allowing the larva to rise and fall in the water. One unusual thing about this creature is the fact that its antennae are adapted to seizing its prey.

Pupa The pupa (size 8mm, approximately) is similar in shape to the chironomid pupa, except instead of having white breathing spiracles on top of the head it has two ear-like appendages which act as a form of stabiliser. It breathes through the leaf-like tails at the end of the abdomen.

Phantom Larva (Price)
Hook 14-16
Silk White
Tail A small tuft of white marabou
Body Flat silver tinsel
Head Two turns of white ostrich herl

Artificials

Flydressers and anglers are sublime optimists, for in tying up imitations of the various stages in this small fly's life we try to achieve transparency in our flies. I think we deliberately ignore the fact that our cunningly concocted artificial has a very opaque hook running through the middle and out of the end. But we pretend it is not there and we (myself included) create artificials which are allegedly transparent. Maybe rose coloured glasses would help!

For the dressing of the larva I have once more utilised the very mobile properties of marabou feather fibre. The merest tweak on the line contorts our very dead artificial into a semblance of life.

Phantom Pupa (Price)
Hook 14-16
Body Flat silver tinsel
Rib Green fluorescent silk overlapped with clear polythene
Thorax One white, one orange/brown ostrich herls wound together

Phantom Midge (Price)
Hook Up eyed 14-16
Body Flat silver tinsel wrapped in polythene
Hackle White cock.

Other Dressings
The first dressing was given by the angling entomologist David Jacques in his book *The Development of Modern Stillwater Fishing*, a work which deals very thoroughly with the Phantom Midge.

Phantom Pupa (D. Jacques)
Hook 15
Silk Black
Body PVC dyed lightly in picric acid over white swan or goose herl or flat silver
Thorax Cinnamon turkey tail or wool of the same colour.
David Jacques wisely has left the larva well alone.
The next dressing comes from the vice of that all round angler and entomologist John Goddard, whose two books on natural trout flies have yet to be equalled.

Phantom Pupa (J. Goddard)
Hook 16
Body White silk
Rib Narrow silver tinsel
Thorax Orange silk
Both the body and thorax are then wrapped in PVC.

Phantom Midge (J. Goddard)
Hook Up eyed 14
Silk Orange
Body Grey condor herl
Rib Olive dyed PVC
Wings White hackle points tied spent
Hackle Honey
This pattern of John Goddard's represents the female Phantom fly as she returns to the water to lay her eggs. He recommends it to be fished in the margins.

Phantom Pupa (C. F. Walker)
Hook 16
Body Nylon
Thorax Ostrich, peacock or condor herl (he gives no specific colour)
Hackle Short hen hackle with fibres at the top cut off

Phantom Midge (C. F. Walker)
Hook 16
Body Pale blue green nylon
Thorax Tying silk (presumably of an olive or brown shade)
Hackle Grizzle cock long in fibre, tied in bunch fore and aft.

Polyvinylchloride
PVC, as it is more normally called, is used quite often in flydressing these days; one of the great advocates of this material is John Goddard, who uses it on many of his very effective imitations, such as the famous PVC nymph, Shrimper and Hatching Midge Pupa. This material was readily available in the high street stores, where it was sold as shower curtaining. And how about the old PVC mackintoshes (whatever happened to them)?

Fishing the Phantom

The Chaoborus hatches out on warm days throughout the fishing season, but it is at its most prolific, or so I have found, during July and August and should be looked for at that time. Check the shady vegetation that surrounds the water where you fish, and look for these tiny, insignificant little flies, that rest 'bottoms up', as it were.

How to fish them? Well, the pupal and larval imitations should be fished on a greased leader with hardly a movement. Moving an inch at a time is in all probability too much; you must fish them with a studied degree of patience, with perhaps your fingers crossed. As for the adult, cast in the vicinity of rising fish and hope that one will be daft enough to take your imitation—if it does then you can uncross your fingers and remove your tongue from your cheek!

The Mosquitoes

ORDER: Diptera
Family: Culicidae

There can be very few of us who have not lain awake at night listening, with a degree of trepidation, to the Stuka-like whine of a marauding mosquito as it flies around the bedroom looking for its supper. The time really to worry is when it switches off its engine; then you know it has homed in, and the inevitable itchy bite will soon follow. The reason why I have included this insect so early in my work is not because I consider it an important source of trout food, but because it is closely allied to the two preceding insects, the Chironomid and Chaoborus midges.

The mosquitoes belong to the family Culicidae, and are divided into two groups—the *Culex*, and *Anopheles*. It is the female of the latter (*Anopheles maculipennis*) that carries the malarial parasite, and it was not so long ago that malaria was prevalent in the British Isles.

In Britain we have thirty different species of mosquito (four *Anopheles* and twenty-six *Culex*). All start life in water—stagnant ponds, quieter parts of rivers, water butts, even in rainwater trapped in the boles of trees. Where they occur in our trout waters they are eaten by the fish. We all know what this little bloodsucker looks like; sufficient to say that it closely resembles the two preceding flies, with the exception of its mouthparts. In the mosquito (female) they are adapted for bloodsucking, and are long and stiletto-like; the males use theirs for sucking nectar in flowers.

Life Cycle

The female lays her eggs on the water surface in boat shaped rafts. These can easily be seen in rain butts if you look closely enough. The eggs hatch out in time on the surface and the infant larvae sink below the surface. The larvae are free swimming, and feed on microscopic food particles suspended in the water, sweeping them in by means of minute hairs. *Culex* mosquitoes breathe at the surface by means of a single siphon; the *Anopheles* does not have such a mechanism but has two spiracles on its abdomen.

The pupa resembles the pupal stages of the Chironomid and the Phantom Midges. It lacks the white breathing spiracles of the 'buzzer', but breathes through two small funnels on its thorax; this thorax has a compartment of air which allows the pupa to hang in the surface film, its abdomen pointing downwards. When it is time to hatch out the abdomen is raised to the horizontal, the skin splits, and the adult fly emerges.

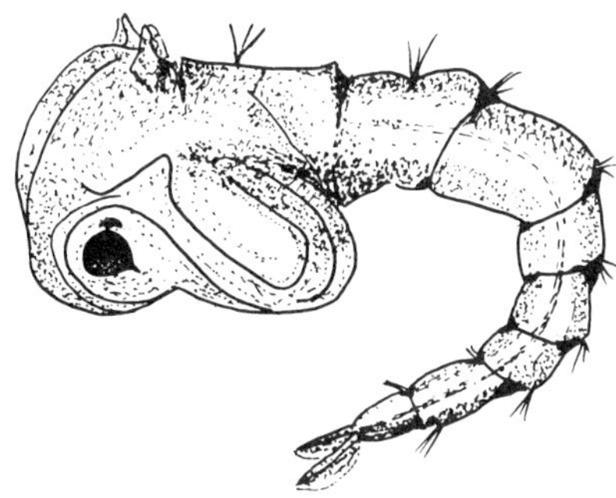

Mosquito pupa

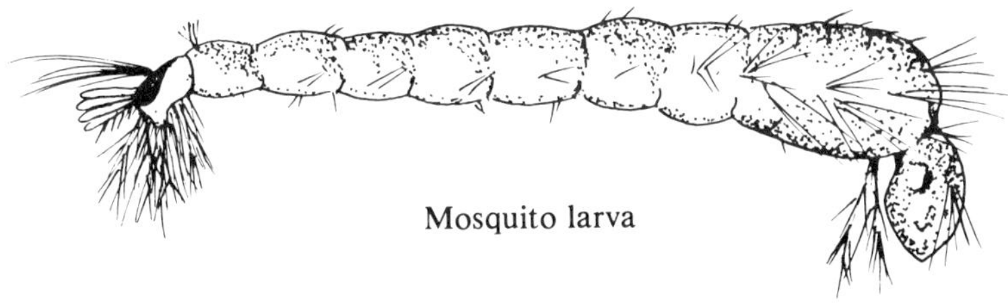

Mosquito larva

Artificials

As far as Great Britain is concerned the mosquito as an artificial trout fly has virtually been ignored. Far more attention has been paid to it by American fly dressers, so I will include some of their patterns to make up for the lack of British versions.

Mosquito Larva (Price)
Hook 12-14
Body Grey silk
Rib Black terylene or silk
Hackle White or blue dun clipped short

Mosquito Pupa (Price)
Hook 14 wide gape
Body Grey silk carried right around the bend, ribbed with black silk
Thorax A full ball of mole or grey rabbit underfur

Mosquito (Price)
Hook 14-16
Body Thinly dubbed grey polypropylene
Rib Black silk
Wings Two small grizzle hackles tied spent
Hackle Blue dun

Terylene
What has been said for nylon can be repeated for this man-made fibre. One use to which I put it is the ribbing of dry fly bodies. I cannot see much point in tying a fly that is meant to float and then ribbing it with heavy metal wire. Most of my dry flies are ribbed with various coloured terylene threads, bought at the local needlework shop.

Fishing the Mosquito

I always consider some flies, and the artificials tied to represent the mosquito are a prime example, to be 'maybe' flies. You know—maybe they will work, and maybe they won't. I must admit, however, that an adult mosquito fly has caught me a number of fish, albeit on a river. Fish the pupa and larva just under the surface on a long, well greased leader. Fish the fly in the same way as the 'buzzers' and phantoms. (If you take my earlier advice, and cross your fingers when fishing the Phantom, cross your legs for this insect and pray to the water gods.) The adult fly should be fished on the surface with no movement at all, especially where surface activity is noted. Mosquitoes hatch out throughout the season, but for most of the time the trout are going to be occupied by more palatable fare. I would tend to fish these flies on those hot summer evenings in shallow sheltered bays, when the trout come in close to see what they have been missing earlier in the day when the clumping waders of daytime anglers disturbed their peace.

Additional Patterns

Mosquito Larva (USA)
Hook 12-18
Tail Grizzle hackle fibres
Body Dark fur overlapped with a stripped peacock quill
Thorax Peacock herl
Feelers Grizzle hackle fibres

Mosquito Pupa (D. Collyer)
Hook 12
Body Stripped peacock quill
Thorax Mole or muskrat fur

Mosquito 1 (USA)
Hook 12-18
Body Dark and light moose mane twisted together
Wings Grizzle hackle points tied upright
Hackle Grizzle
Tail Grizzle hackle fibres (this is an aid to floating—the natural has no tail)

Mosquito 2 (USA)
Hook 12-14
Body Stripped peacock quill
Hackle Dark grizzle
Wings Dark grizzle hackle points
Tail Grizzle hackle fibres (again a floating aid)

The Sedges

ORDER: Trichoptera

The Sedge flies belong to the order of insects known as Trichoptera, meaning hairy winged. These insects appear to be closely related to moths, but whereas a moth's wing is covered in minute scales, a sedge fly's is covered with tiny hairs. We have in Britain about 197 different species; worldwide there are approximately 1000 identified species.

In this country the angler has always referred to this insect as a Sedge, presumably from its habit of resting in the sedges that surround many of the rivers and lakes where the creatures hatch. In the United States it is invariably referred to as the Caddis fly in all three stages of its life. We Brits only use the term caddis when describing the larval stage. In parts of Britain the insect is referred to by some as a Sedge Moth.

On most waters the sedge is an important source of food for the trout, more important than members of the Mayfly family because there are usually more of them, both in numbers and in species.

The insect certainly resembles the moths in many ways, not only from the fact that its physical appearance is similar but also from its habit of flying at dusk or at night. In my moth trap, amongst the many species of moths, I always find a number of sedges. I am at least half a mile from the nearest water, and have so far trapped four species. These flies have four wings which they carry roof-like over their backs. They are in the main sombre coloured insects possessing long antennae. They vary in size from about 15mm to 27mm, depending of course on the individual species.

Life Cycle

The sedge undergoes a complete metamorphosis, thus

Egg → Larva → Pupa → Adult

Depending on the species, the eggs are laid in a variety of different ways. In some the female flies over the water, dipping the tip of her abdomen beneath the surface to wash off the eggs. In others the eggs are dropped in a small green ball from just above the surface. Some species lay their eggs on overhanging vegetation, and yet other females go below the surface to deposit their eggs on the bottom.

Larvae In the Middle Ages most of the country's population lived in isolated communities; roads were practically non-existent. These communities were served by an itinerant and varied band of pedlars, and one such pedlar sold a wide range of braids, yarns and ribbons. One yarn was called caddis or caddice, an inexpensive material. In order to display his wares the pedlar sewed pieces of the material to his clothing, and he became known as a Caddice or Caddis man. Supposedly this is how the larval stage of the sedge fly obtained its name, for it was noticed that this little creature bedecked itself in a wide variety of debris as it crawled along the river or lake bottom, and so it became the Caddis Grub.

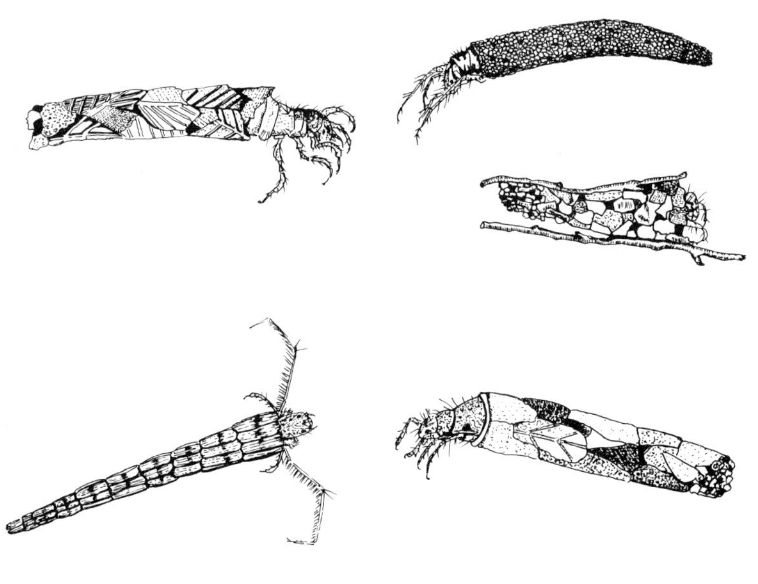

Caddis cases

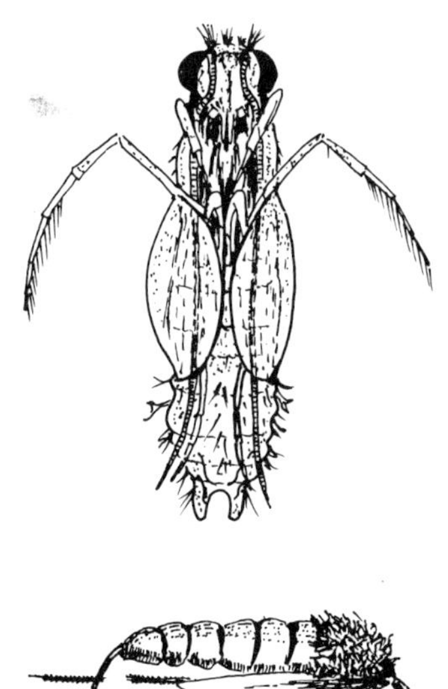

Sedge pupa—natural and imitation

Caddis larva out of case

Not all larvae build these intricate shelters. Some species create webs or funnels, while others are free swimming. It is the larva that builds its portable shelter, however, that attracts most attention and wonder. These portable shelters serve three purposes: protection, camouflage, and (in rivers) ballast, preventing the insect from being carried away by the current. In still water many species utilise much lighter materials for the opposite reason—they are not concerned with currents, they require mobility.

When it is time to pupate the larva seals up the end of its home with a fine porous silk membrane, sometimes attaching itself thereby to stones.

Pupa When it is time to hatch out the pupa breaks free of its pupal prison and swims quickly to the surface in a thin caul-like membrane. On reaching the surface it soon disposes of this membrane, and either runs helter skelter (in angling parlance, 'skitters') across the surface or takes to the wing immediately its wings are dry. Some species swim under water to the nearest vegetation and crawl up the stems to hatch out.

This hatching pupa·is perhaps the most killing stage to imitate with our artificials.

Adult Sedges Some adult species only live for a week or two. Others, when first hatched, lack the ability to mate, and hide up in vegetation for a month or so until their genitalia are formed completely. The sedges are in evidence right through the fishing season, but it is from June onwards that they hatch out in large numbers, attracting the attention of the trout.

We have in this country one species of Sedge fly that is wholly terrestrial (*Enoicyla pusilla*). Like its aquatic cousins it builds itself a case of sand. It is only found in the Forest of Wyre near Worcester and feeds on dead oak leaves.

The trout feeds on the caddis or sedge fly in all its three stages. I have taken many trout early in the season stuffed to the gills with caddis larvae—later in the season it is the pupa and adult that interest our quarry.

Sedge Larvae Patterns

The first pattern is one I tied up many years ago and found to be reasonably effective in both its forms. I used clipped deer's hair as its main constituent, not for its buoyant properties, for I weighted the fly, but rather for the fact that it could be easily clipped into shape; smooth to imitate the sandy type of case, and shaggy to imitate the caddis that creates its home out of twigs and other debris.

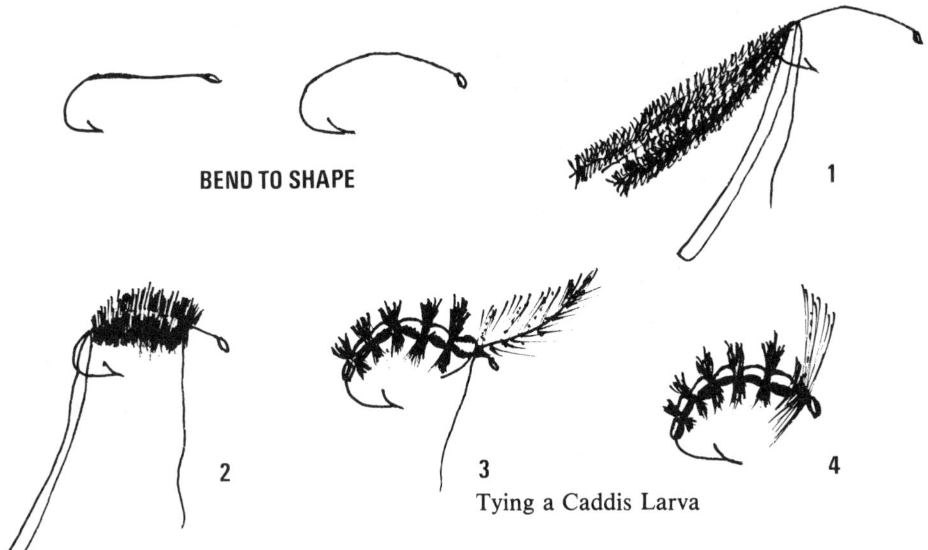

Tying a Caddis Larva

Deer Hair Caddis
Hook L/S 14-8
Silk Brown or olive
Body Clipped deer hair
Hackle Sparse (colour immaterial)
As far as the hackle is concerned I like to exaggerate slightly to give the artificial a degree of mobility; I find that this is more attractive to the trout than a more natural short hackle.

The next two I published in *Rough Stream Trout Flies* (A. & C. Black). I do not claim they are unique. I have been in the business too long, and realise that very few anglers or flydressers create new patterns; there is little new under the sun, and this applies to the invention of fishing flies in particular. When I

S. D. (Taff) Price, *Rough Stream Trout Flies*, A. & C. Black.

first started tying I, like all beginners, invented hundreds only to find I was in some cases a hundred years too late. Never mind, here are a couple of patterns for someone else to reinvent. On the face of it these two flies, though appearing very natural, have a touch of over realism about them. They can look too good, if you know what I mean, but I can assure you they work—more precisely, they have caught fish for me.

A third pattern of a similar nature—and we can term this one the Sand Caddis—is tied to represent the caddis larvae that build their cases like miniature cornucopia.

Twig or Debris Caddis

Hook L/S 14-8 (weight hook towards the head)
Silk Green
Body Either green or yellow floss

Stick to this floss, by means of adhesive, small cut portions of either green or natural raffia, or both, with, if desired, a small soft twig or two. Finish off with a small hackle to represent legs.

Gravel or Pebble Caddis

Exactly the same procedure as the preceding pattern, but instead of raffia, etc., stick on (I use *Araldite* for this one) some minute pebbles or gravel and allow to dry. (There is no need to weight this one. It sinks like the proverbial brick.) This last pattern I used on the opening of the 1977 season. It took seven brown trout, and I lost many others. But be warned—it is a bit of a cow to cast!

Sand Caddis

As for the other two, but this time taper the body more and roll the adhesive covered fly in fine sand, allow to dry, then put on the small hackle.

In the last couple of years the material 'latex' or dental dam has become popular, especially with many American dressers. The following pattern I have developed to imitate the free swimming larvae of the *Rhyacophilia* or *Hydropsyche* species; this uses latex more or less as a ribbing medium, rather than as a body.

Latex Caddis

Hook L/S 14-8, weighted
Body Olive green ostrich herl at least four strands
Rib Pale yellow dental dam (the rib is wound on leaving the herl to protrude as breathing filaments). The herl is clipped off top and bottom
Hackle Sparse olive but colour not important

Latex

Sometimes referred to as dental latex, rubber dam or dental dam, this thin sheet rubber is used by dentists to mask off teeth during their diabolical operations. The recent popularity of this material, stemming originally from the United States, has given rise to a host of stonefly creeper, sedge pupa and larval patterns. I have tied some myself, and all the flies look extremely realistic. But, a word of warning. On its own latex, though

The next pattern can well be described as the forerunner of the Latex Caddis above; it was given to me by Gerry Emberson who fishes the southern reservoirs with more than a marked degree of success. This Stick Caddis Fly, though it catches fish, has one drawback. It is somewhat delicate in construction. I for one do not mind the short life of a fly, if it has proved its worth and finally disintegrates in the teeth of a caught trout.

There are many patterns called simply Stick Flies. All of them are very similar in make up, and the second pattern above is just one of them. Simple in construction, they are extremely good trout deceivers.

Stick Fly

Stick Caddis Fly (Emberson)
Hook L/S 10-8
Body Floss silk various colours such as olive, yellow, orange
Rib Ostrich herl contrasting with the body colour
Legs Three strands of cock pheasant tail fibres tied flat on the top of the hook and clipped to size. A head of peacock herl can be added for a more natural effect.

Stick Fly
Hook L/S 14-8
Silk Brown or black (not important)
Body Thinly dressed peacock herl
Rib Gold or silver tinsel
The addition of a fluorescent, thoracic segment enhances the fly
Hackle Brown (again colour matters little)

Before leaving the general larval patterns I must mention one caddis larva that I have collected in Darwell reservoir. It belongs to a species that has no angling name; its entomological name is *Triaenodes bicolor*. This larva is one of the few caddis that swim with their cases rather than crawl like most others. Its legs are developed for swimming, and I have seen evidence of this creature in the stomach contents of trout. It greatly resembles a thin tapering reed, but on closer examination you will find that it is an ingenious creation of spiral leaf formation. The following pattern imitates this creature; it should be fished in the vicinity of weed beds, with a fairly fast recovery, for the

giving a lifelike appearance, is completely dead. You may as well fish with those plastic moulded abominations that catch only anglers. Latex must be used in conjunction with mobile fur or hackle.

There are two ways of applying latex strip to form the abdomen of an insect. Over a wool or fur, and applied tightly, the resultant body is plump and non-segmented. An alternative method is to wind the strip slightly overlapping each previous turn, and not so tight. The result is a segmented effect imitative of stonefly nymphs.

The use of rubber for the bodies of flies is anything but new; many prewar mayfly patterns had rubber incorporated in their body construction. Another form of rubber that can be used to good effect is the material sold by florists' sundriesmen for binding the stems of flowers and plants. This material is not quite as strong as dental dam, but has more mouldable qualities.

Dental dam is now available from the usual flydressing houses, but if you can persuade your wives to purchase light yellow rubber gloves for kitchen use (instead of the red ones) then you will have a limitless supply of flytying latex.

natural insect moves at a fair rate of knots once it gets going, when caught in the open water away from the camouflaging weed beds.

For comparison's sake the next two patterns are American in origin, and certainly worth a try on our trout.

Triaenodes Caddis
Hook L/S 10
Body Thin strip of green raffia (varnished)
Rib Black silk
Hackle Sparse blue dun
For comparison's sake the next two patterns are American in origin, and certainly worth a try on our trout.

Little Gray Caddis
Hook L/S 8-10
Silk Black
Body Thinly spun gray fur
Ribbing Thin gold wire and peacock herl
Hackle One turn of black hen short in fibre

Breadcrust (USA)
Hook L/S 10
Silk Black
Body Orange or green floss or wool
Rib Stripped hackle stem from a brown hackle
Hackle Grizzle

Fishing the Caddis Larva

All the above patterns, with the exception of the Triaenodes Caddis, should be fished as close to the bottom as possible, for that is where the natural creature is to be found. To do this either use a sinking line, sink tip or (if using a floating fly line) a long leader. Fish with a slow retrieve; try and imagine your fly as the real thing; imagine it exploring the bed of the lake or reservoir and endeavour to impart the same movement to it. Fishing is as fickle as a woman, for I know certain anglers who make killing catches by stripping 'stick flies' fast just under the surface. If that's what turns you on, then please yourself. You can safely fish these caddis larvae imitations right through the season, for the insect is always in evidence. Early in the fishing calendar I have found to be the best time, but later in the year when there is no certain surface activity resort to using one of the aforementioned patterns. It could work the oracle.

General Pupal Patterns

Of the three stages in the life cycle of the Trichoptera, I believe the pupal stage to be the most effective from a fish catching point of view. Quite often the trout appear to be rising to the adult insect, but what is happening is that they are chasing the fast rising pupae to the surface and, carried on by their own momentum, are breaking the surface. As far as artificials are concerned there are many to choose from, from Dr Bell's

Amber Nymph to modern latex concoctions. The first pattern I can hardly call original in conception, for I have based it on well-tried American styles utilising latex as the body medium.

Sedge Pupa (Price)
Hook From normal shank 12 to L/S 10
Underbody Floss silk or wool
Abdomen Cream latex marked with felt pen
Thorax Orange or green seal's fur mixed with hare's fur
Wings Two short grey duck, or mottled hen, tied either side of the thorax
Legs Grouse hackle fibres
Antennae Brown mallard fibres or in larger sizes cock pheasant tail fibres

Tying notes
1. Tie in latex and underbody material.
2. Wind on underbody.
3. Wind on latex in overlapping turns.
4. Tie in wings.
5. Dub on thorax fur, and form thorax.
6. Tie in antennae fibres, and grouse hackle.
7. Wind on hackle, slope back, and whip finish.

Tying an imitation latex sedge pupa

The next pattern is the notable Amber Nymph created by Dr Bell, of Blagdon fame. It does not greatly resemble a sedge pupa, but its overall shape and coloration stamps it a sedge pupa imitation. Dr Bell had two such patterns, in fact, so to keep the record straight I'll give them both. He recommended the large Amber Nymph to be fished in May and June, the smaller version in July.

Large Amber Nymph (Dr. Bell)
Hook 10 (I've found size 8 also to be effective)
Body Yellow floss or seal's fur
Thorax Brown floss or seal's fur
Back (Wing case) Grey brown feather
Hackle Pale honey dun or light ginger

Small Amber Nymph (Dr. Bell)
Hook 12
Body Yellow floss or seal's fur
Thorax Hot orange silk or fur
Back (Wing case) Grey brown feather
Hackle Honey dun or light ginger

Now let us get back to more modern times. An imitation tied to represent a wide number of species is the next, devised by John Goddard, author of *Trout Fly Recognition* and *Trout Flies of Stillwater.* John recommends the fly to be fished in medium to shallow water and activated or retrieved slowly.

John Goddard, *Trout Fly Recognition*, A. & C. Black.

John Goddard, *Trout Flies of Stillwater*, A. & C. Black.

Judging from the number of this pattern I am asked to tie in the course of the year it must be one of the most popular sedge pupa patterns in use today.

Sedge Pupa (J. Goddard)
Hook L/S wide gape 10-12
Silk Brown
Body Various colours, cream, olive green, orange
Rib Silver lurex
Thorax Brown condor herl
Wing case Light condor herl (cinnamon turkey will suffice)
Hackle Up to two turns of a honey hackle

Goddard's Sedge Pupa

Richard Walker needs no introduction from me. Whether you agree with everything he writes or not you have to admit that this controversial figure has stamped his mark on modern angling in this country. One of his 'bêtes noires' is the incorrect dressings given for his flies in some books (mine included) so rather than be castigated by him for the rest of my life **(the hackle on the Sweeny Todd is not magenta)** I shall endeavour to get the next pattern of his correct. Richard's best known sedge pupa pattern is the Longhorns, tied in two colours. The original used ostrich herl for the body, although he intimated to me subsequently that he now ties the patterns with wool instead.

Longhorns (R. Walker)
Hook L/S 10-12
Body Ostrich herl (colours: amber, sea green)
Rib Gold thread
Thorax Chestnut ostrich
Hackle Brown partridge
Horns Cock pheasant tail fibres

Longhorns

To imitate the smaller sedges that inhabit most still waters, Richard Walker has devised a dressing to imitate their pupae. He maintains that the following pupal dressing (which he has named the Shorthorns) is an effective pattern when such sedges as the Brown and Black Silverhorn and the Grouse Wing sedges are in evidence.

Very occasionally one angler becomes associated with a river. Skues on the Itchen, perhaps Izaak Walton on the Dove, and so forth. When we think of the Usk then one man springs immediately to mind and that is the late Lionel Sweet. To see Lionel Sweet cast with a centre pin reel was to see perfection; he could cast further and with greater accuracy than many other anglers equipped with fixed spool reels. The following pattern is a fly he designed primarily for use on still waters such as Chew and Blagdon. (It also takes trout on rivers.) An interesting feature of this fly is the double rib, the effect of which is similar to that which a signwriter seeks to achieve when he wishes to emphasise a letter. He introduces a shadow. This Lionel has done with his Amber Nymph. He followed a gold rib with one of black thread, thus emphasising the feature.

Shorthorns (R. Walker)

Hook 12-14
Silk Brown
Abdomen Dark brown olive feather fibre
Rib Yellow tying silk
Thorax Greeny yellow fluorescent wool, ball shaped (orange as alternative)
Wing case Black lurex
Hackles Brown partridge tied in two bunches at each side and sloping backwards

Sweet's Amber Nymph (L. Sweet)

Hook 10-12, weighted
Silk Black or brown
Tail Wisp or two of red (natural) hen
Body Amber floss
Rib Gold oval followed by black thread tight up against the gold
Thorax Amber floss
Hackle Soft natural red hen

Yorkshire Fly Body Material

The firm of Mackenzie-Philps Ltd market a very good fly body material made from man-made fibre; it is sold as a yarn which can be wound on its own, although the fibres can be pulled out and dubbed in the usual way. It comes in a full range of colours.

This next pattern was created by Peter Mackenzie-Philps of the Yorkshire tackle firm of the same name, which is well known for the ingenious design of hook they invented (as well as a host of aids and materials for the creative fly dresser). This fly of Peter's, which he calls Sweetie's Sedge, attempts to imitate the hatching sedge in the surface film, and I can do no better than to repeat the essentials from his letter to me.

Sweetie's Sedge
(C.P.B. Mackenzie-Philps)

Hook	10 or 12, light on the the wire
Silk	Olive (Pearsall's colour 16)
Body	Pig body fur or seal's fur, almost apple green. (The colour needs to be bright—the trout can be very fussy in total darkness)
Wing case	Slips of starling on grey duck, tied in below and brought sloping upwards and forwards to be tied in behind the head
Thorax	Flybody fur or seal's fur, sedge brown or nut brown with a tinge of red, tied bulky
Legs	Photodyed blue dun hackle fibres, bearded in below, and extending to the point of the hook
Antennae	Two fibres from a mallard shoulder feather, tied long, and laid over the body. Care must be taken not to break these off when tying the fly on the leader—they seem to be critical to the trout's recognition of the silhouette.

Tying notes Run the silk down the shank of the hook, to a point well round the bend. Dub the fur for the body, and bring the body forward to a point half-way between bend and eye. Half hitch, and then turn the hook upside down in the vice. Tie in the hook, and by their tips. Now dub the thorax, so that it is noticeably bulkier than the body. Beard in the hackle, and half hitch. Turn the hook upright in the vice. Bring the wing cases forward to the front of the thorax, and tie in on top, just behind the head. It helps to do this one at a time. And I have to use tweezers or dubbing needle to bring them forward. Tie in the two mallard fibres at the head, and make sure that they are at least twice the length of the body. Finish the head.

Fishing notes It is important that any faults in line or leader. under the surface film. When there is a rise going on the fish are all close under the surface, and every fish I have caught on this pattern has been a good confident rise with a big swirl. I do not think that fish can see below themselves, so it is no good fishing this pattern sunk. I make sure that my line is clean, and I grease the leader to within two feet of the fly. I have tried oiling the fly, but this seems to reduce its appeal. I choose the eastern shore, and fish towards the last of the light in the sky so that I can see the light reflected on the surface. It helps greatly if I kneel down, as the surface looks much lighter then, and short casts are all that is necessary—most of my fish have been caught within two rod-lengths, in about six feet of water. I use only one fly, as there is nothing more annoying than to hear that swishing which denotes a tangle when the fish are going daft and you realise that there is perhaps only another ten minutes of fishing light left! Move the fly ONE INCH at a time, and this only to draw the attention of the trout. Many fish have taken the pupa as soon as it dipped into the ring of a rise and before I could even straighten out any faults in line or leader.

Before coming into the fishing tackle trade Derek Bradbury of Cheshire was an artist, witness his beautifully illustrated articles in the *International Flyfisher* magazine. A longstanding member of the Fly Dresser's Guild (his artistic talents also grace the cover of the Guild magazine), Derek Bradbury has created some very effective flies, one of which he sent me for my first book on lures. His version of the sedge pupa is his Amber Nymph, the dressing of which is as follows.

Amber Nymph (D. Bradbury)
Hook 10
Silk Black
Body Rear two thirds amber seal's fur, final third dark brown ostrich
Rib Orange lurex (over seal's fur only)
Back Dark mottled turkey tied at tail and taken over the back
Hackle One turn of brown partridge
Paddles Two fibres from a Golden Pheasant tail pointing rearwards, possibly imitates the antennae of the sedge pupa.

Fishing Sedge Pupae

I always fish the pupal imitation of the sedge on as long a leader as possible, using a floating line. The fly is allowed to sink, and retrieved in long steady pulls in order to imitate the natural insect as it swims up to the surface. Sometimes it is effective to fish the imitation in the surface film, and to simulate the struggles of an insect trapped in the film as it endeavours to hatch.

Adult Sedge Imitations

A number of artificials are given later in this chapter under each specific insect. There are, however, a number of flies designed to imitate the general conception of a sedge fly without being too specific. In this category we have those bushy flies known to reservoir anglers as Wake Flies. The following Wake Flies are simple to construct and are as effective as any.

Brown Wake Fly
Hook L/S 10 or Mayfly hook 10
Body Palmered brown hackles
Wing Either set together streamer fashion, or tied in a 'V' back to back
Front hackle Two brown cock hackles

Others are as follows:
Black Wake Fly
As for Brown but using black cock hackles
Furnace Wake Fly
As above using furnace (Coch y bonddhu)
Badger Wake Fly
As above using Badger cock
Grizzle Wake Fly
As above using a well marked grizzle.

The last two also make very good moth patterns for dusk fishing, when the moths are flitting and the big browns eagerly wait to take them as they foolishly fall on the surface. All the above patterns should be well treated with silicone floatant, cast out and drawn back to the shore causing a wake/disturbance on the water. (Dry Muddler Minnows should be fished in the same way, for the Muddler is truly one of those multipurpose flies that can be whatever the trout want it to be. In silhouette it can with very little imagination resemble a sedge.)

The following selection of general sedge fly patterns are the work of anglers and flydressers from various parts of the country.

The first pattern is the combined brainchild of John Goddard and his collaborator Cliff Henry. It is a sedge pattern to imitate the lighter sedges, and has a unique wing made of clipped deer's hair. This pattern is a little difficult to tie, so I will give instructions as well as the dressing.

The G & H Sedge
(Goddard/Henry)

Hook	L/S 8-10
Silk	Green
Body	Clipped deer's hair (this is really the wing)
Underbody	Dark green seal's fur (see dressing notes)
Hackle	Rusty dun cock (two)
Antennae	Butts of the two hackles

Dressing procedure
Tie in a length of tying silk in at the tail and leave, then spin on deer hair in the usual manner (as for Muddler head), leaving enough room at the head of the fly for the hackles. After spinning on the hair, clip to a sedge wing shape, i.e. wide at the tail end, tapering down to the eye. Now reverse the hook in the vice and onto the tying silk spin the green seal's fur. Take this thread along the body and tie in (this stage can be substituted) with two strands of green or orange wool. (I find this easier than the spun seal's fur.) Replace the hook the right way in the vice and tie in two hackles; do not clip the stems—these can be the antennae; wind on hackles and finish off the fly.

I was given the following pattern by Chris Padley of Frant near Tunbridge Wells and it is a fly that proves very successful on Weirwood Reservoir.

Orange Sedge (C. Padley)

Hook	12-14 up eye
Body	Hare's ear
Rib	Flat gold lurex or tinsel
Wing	Rolled partridge tied over the back
Hackle	Hot Orange

Why this pattern utilising bright hot orange hackle works I cannot begin to guess but it does.

Whenever I catch and photograph a sedge and have difficulty in identifying the same I place it in a small plastic container and pop it in the post to David Jacques to do the hard graft for me. His knowledge of the Order Trichoptera and anglers' insects generally is second to none. The following broad spectrum pattern is one of his.

Pale Sedge (D. Jacques)

Hook	14-10
Body	Cinnamon turkey
Rib	Gold twist
Body hackle	Ginger palmered cock
Wing	Rolled hen pheasant wing fibres tied flat over the body
Hackle	Ginger cock

Standard Sedge (T. Thomas)

Hook	All sizes
Body	Cock pheasant tail fibres
Hackle	Palmered ginger cock
Wing	Grey deer hair tied by the tip and splayed out
Head hackle	Ginger cock

Cree Sedge (R. Masters)

Hook	16-14
Silk	White
Body	Cree hackle palmered then clipped
Wing	Natural red rock flat over the back and trimmed
Hackle	Cree cock

In the last few years there has been a quiet revolution in the world of fly dressing. One of the reasons for this has been the influence of many books on the subject from over the Atlantic ocean. Books such as *Selective Trout* by Swisher and Richards (with their No-Hackle and Parachute style flies); the monumental work *Nymphs* by Ernest Schwiebert; *The Caddis and the Angler* by Solomon and Leiser, are but three of many books that are now available to the British angler. The inventive skills of such American fly dressers as Dave Whitlock and Poul Jorgensen appear in many of these publications from the USA.

The following patterns (traditional and modern) from the States are well worth dressing and trying on our reservoirs. Most American caddis patterns were designed for river use, as are most of their streamer and bucktail flies.

Doug Swisher & Carl Richards, *Selective Trout*, Crown (New York).

Ernest Schwiebert, *Nymphs*, Winchester Press (New York).

Larry Solomon and Eric Leiser, *The Caddis and the Angler*, Stackpole (New York).

Delta Wing Caddis

Hook	10-8
Silk	Olive
Body	Olive fur dubbing
Wing	Grey hackle tips tied delta fashion swept back
Hackle	Brown

Bucktail Caddis

Hook	10-16
Silk	Brown
Tail	Short brown deer's hair (the natural caddis of course has no tail)
Body	Tan polypropylene
Hackle	Palmered brown cock
Wing	Natural brown deer hair
Head hackle	Brown

Light Colorado King

Hook	10-18
Silk	Black
Tail	Peccary fibre tied long and at an angle (substitute moose main fibres)
Body	Yellow dyed rabbit
Body hackle	Grizzle
Wing	Deer's hair longer than hook shank, tied flat along the body

Deschutes Caddis

Hook	8-14
Silk	Brown
Tail	Ginger hackle fibres
Body	Yellow raffia
Body hackle	Palmered ginger cock
Wing	Elk or deer hair tied flat over body and sparse
Hackle	Brown cock

Dark Caddis

Hook	All sizes
Silk	Yellow
Tail	Yellow hackle fibres
Body	Yellow wool or polypropylene dubbing
Body hackle	Bright yellow (golden)
Wing	Dark.bucktail

It is interesting to note that most American Sedge (or, as they term them, Caddis) flies are often adorned with tails. This is probably an aid to the floating potential of the fly, for the natural insect has no tails.

Fishing the Adult Sedge

Sedge flies, as I have already stated, hatch throughout the fishing season, but they are most in evidence from the end of May onwards. Fish your artificials on a well greased leader, waking them back to shore like the 'skittering' natural. Another method is to cast out the fly and retrieve in short jerks with pauses in between.

Selected Sedges

THE CINNAMON SEDGE (*Limnephilus lunatus*)

This is one of the angler's most popular sedges. Although *L. lunatus* is considered to be the Cinnamon Sedge, other members of the *Limnephilidae* of similar size and coloration are also taken as the Cinnamon. *L. lunatus* is a well distributed species, found in both still and running water, and it has received the attention of many flydressers.

It is sometimes thought to have obtained its name from its overall coloration, but this is not so. It is known as the Cinnamon Sedge because it is supposed to emit a smell of cinnamon. David Jacques, an expert on Trichoptera, thinks it smells more like geranium. Me, whenever I have tried to smell one of these insects all I can detect is the leftover scent of my pipe tobacco. (The day I give up smoking there will be something else new to look forward to . . . I'll be able to smell sedges.)

Larva: L. lunatus is a case building species, constructing its mobile home usually out of a wide variety of vegetable and leaf debris laid longitudinally, and often including gravel, sand and snail shells. (It is possible to identify many species of caddis from the type and materials used in the case construction, always assuming the usual materials preferred by the insect are available. If they are not then it will utilise whatever is at hand, even taking over the discarded cases of other species, so clearly there are no hard or fast rules.)

General Identification
Size: approx. 15mm
Colour: Light brown
Antennae: Stout but as long as wing
Wing: The pale cinnamon wings have a few dark patches, one at the top towards the rear of the fly, a larger half moon marking followed by a clear patch at the end of the wing. It also has several paler patches in the centre of the wing.
Body: Sometimes brown, but on occasions green

As with all sedges positive confirmation of identification can only be carried out with the aid of the microscope.

Adult Emergence It is first on the wing in June, but it flies in large numbers at dusk in late summer and autumn.

Artificials

Cinnamon Sedge (Price)

Hook	12 or L/S 14
Silk	Brown
Body	Light brown polypropylene
Body hackle	Ginger cock
Rib	Dark brown silk
Wing	Ginger hackle fibres clipped to shape extending beyond the hook shank
Hackle	Ginger (two)—stalks can be left to form antennae

As an alternative tie up the same fly but use green polypropylene dubbing.

Alternative Pattern

Cinnamon Sedge (R. Wooley)

Hook	12-14
Body	Cinnamon turkey wing fibre
Rib	Gold wire
Hackle	Palmered ginger cock
Wing	Landrail (substitute with light brown hen wing)
Head hackle	Ginger

The next pattern was devised by Col J. T. Lane and given in his book *Lake and Loch Fishing for Trout* (Seeley Service). For the bodies of his sedge fly patterns Col Lane preferred a clipped hackle. Just in passing, those readers who have not read his book would be well advised to do so. Although much of the book has been surpassed by more modern approaches to still-water fishing, many of his ideas and observations have stood the test of time, and the whole book is laced with a gentle brand of good humour.

Cinnamon Sedge (J. T. Lane)

Hook	12
Silk	Golden olive
Tail	Ginger cock hackle fibres (to aid floating)
Body	Ginger cock hackle
Wing	Pale ginger cock hackle fibres
Thorax	Ginger cock hackle
Head hackle	Ginger cock

One of the great keepers of the river Test provides the next fly. William Lunn is probably better known for his very effective Lunn's Particular, a spent spinner pattern, but his cinnamon sedge is also worth a try. The only fly of his that I am not keen on is his Welshman's Button with the centre stripe of yellow—that creation looks rather more like a wasp or hoverfly than a sedge. The last pattern for the cinnamon sedge I shall give was published in *Chalk Stream Flies* (A. & C. Black) by C. F. Walker. Like Lunn's pattern, C. F. Walker's imitation is of the Cinnamon sedge that appears to have a green body.

Joscelyn T. Lane, *Lake and Loch Fishing for Trout*, Seeley Service.

C. F. Walker, *Chalk Stream Flies*, A. & C. Black.

Cinnamon Sedge (W. Lunn)

Hook	14
Silk	Orange
Body	Swan fibres dyed a light greeny yellow
Rib	Gold twist
Body hackle	Buff Orpington cock hackle
Wing	Well mottled cock pheasant wing dyed cinnamon
Hackle	One ginger and one buff Orpington cock

Cinnamon Sedge (C. F. Walker)

Hook	12
Body	Condor quill dyed dull yellow green
Body hackle	Ginger cock
Wing	Brown mottled hen
Hackle	Two ginger cock hackles

Large Cinnamon Sedge (*Stenophylax sequax*, or alternatively *Potamophylax latipennis*).

Both these flies are large, broadwinged insects, often nocturnal in habit. Both belong to the family *Limnephilidae.*

Larvae

S. sequax: Case making. Straight case built of small pebbles at the head end, fine grains of sand toward the rear of the case.

P. latipennis: Case making. Slightly curved case, about 20-22mm long, comprising small flattened stones and large grains of sand.

The adult insects do not have the dark markings of *L. lunatus*, but have small lighter markings or patches on the wings.

Adult Emergence *Stenophylax sequax* appears from the end of May through to November. *P. latipennis* is on the wing from the end of May to the beginning of October.

Artificials Larger sizes of the smaller Cinnamon Sedge will suffice. As far as overall size and coloration of the natural is concerned there is very little to choose between them and the Caperer (*Halesus digitatus* or *H. radiatus*), so patterns tied to represent this insect will suffice to imitate the Large Cinnamon Sedge.

The Mottled Sedge (*Glyphotaelius pellucidus*)

This is quite a large sedge, being somewhat larger than the more common Cinnamon Sedge. On some waters it hatches out in large numbers and may well be an important source of food for the trout on such waters. The most distinctive feature of this fly is its wing; apart from the heavy mottling, from whence it gets its common name, the configuration is markedly scalloped at the rear, making it quite different from other sedges.

Larva Case making. The larva constructs its case almost wholly from dead leaf matter; it prefers small whole leaves, especially those of the Hawthorn. Larger leaves are cut almost in circles. The larva is very well disguised in its case, for it has a leaf mantle over the entrance, hiding the larva completely.

Adult Emergence From June to October.

Artificials I will not pretend that this is a surefire certainty as far as flies go, but it may well prove useful to those on whose water large numbers of this sedge hatch out. I must admit that on the water where I fish in Kent I have only ever seen one adult insect in all the years I have fished there.

Mottled Sedge (Price)

Hook	10
Body	Pale ginger hare's fur mixed with green seal's fur
Rib	Yellow silk
Wing	Cock pheasant wing extending beyond the hook, and clipped to shape. Further markings can be added with a felt tipped pen if so desired
Hackle	Ginger cock (two)

Mottled Sedge (J. T. Lane)

Hook	10
Silk	Golden olive
Tail	Golden olive cock hackle fibres
Body	Red cock hackle clipped to shape
Wing	Rusty dun cock hackle fibres
Thorax	Red cock hackle
Leg hackle	Red cock

Col J. T. Lane was one of the few angling writers to recognise the worth of the Mottled Sedge. His sedge, which he refers to as the Marbled Sedge, is a different species from *G. pellucidus*, and represents *Linnephilus marmoratus*. Nowadays we would probably term it a Large Cinnamon. In order to give the reader an alternative choice of pattern I include Lane's dressing above.

THE GROUSE WING SEDGE (*Mystacides longicornis*)

One of the most common sedge flies found on still water, and one which hatches out in very large numbers, it can be seen in the late afternoon flying in clouds just above the water surface. As its common name suggests its wing coloration is very similar in colour and markings to a slip of Grouse wing feather. Very occasionally a brood hatches with no dark markings whatsoever, making identification very difficult. Another distinguishing feature is the ultra-long antennae.

Polypropylene

There are two forms of this man-made product that can be of use to the flydresser. One is the usual fibred form purchased from the flydressing suppliers which makes extremely good dubbing, an excellent substitute for seal's fur. The sales splurge for this product tells us that it has a specific gravity of 0.95, lighter than water, and is therefore a floating material. (Don't be fooled—the hook we are going to wrap it around is certainly going to be heavier than water.) This material comes in a wide range of colours. It is easily applied, by wrapping or by teasing out and dubbing on the thread in the usual way, and it does not have that awkward inbuilt spring that seal's fur has. It is much easier to apply. Nowadays it is sold both as a yarn or as a dubbing mixture.

The second form in which one can find polypropylene is in its hard state, as a parcelling twine. It is the modern substitute for the old hairy string which was made from sisal. Polypropylene, being that much cheaper, is taking over in the packaging and post rooms. This form of twine can be easily split, and in flydressing it makes a useful winging medium or body overwrap. It is usually colourless, although coloured versions are occasionally obtainable.

Larva Case making. The case is slightly curved and constructed entirely out of fine grains of sand.

Adult Emergence End of May to September

Artificials It would be extremely foolish of a flydresser to ignore the fact that the wing of the fly resembles the markings and colour of a slip of grouse wing feather, so it is only natural that this medium appears in most dressings of this fly.

Grouse Wing Sedge (Price)
Hook 14
Body Brown polyproplyene or fly body fur
Wing Thin strips of grouse wing extending beyond the hook
Hackle Dark ginger
Antennae Dark mallard fibres

The next pattern is a variation of the above pattern, and differs only in the wing construction. For this artificial I use the tip of a grouse body feather, which is lacquered and allowed to dry before tying in. The first flat-winged sedge I came across was a German sedge imitation.

Grouse Wing Sedge
(Flatwing style)
As above pattern but using body feather just described.

Courtney Williams, in *A Dictionary of Trout Flies* (A. & C. Black) does not give a specific dressing for this insect, which he considers of no value. I would tend to disagree, for if the trout ignore the adult insect they certainly would not turn down the succulent pupa. In *Reservoir and Lake Flies* by John Veniard, Richard Walker gives an effective pattern, as follows:

A. Courtney Williams, *A Dictionary of Trout Flies* (edited by T. Donald Overfield), A. & C. Black.

John Veniard, *Reservoir and Lake Flies*, A. & C. Black.

Grouse Wing Sedge (R. Walker)
Hook L/S 12
Silk Black
Tag White fluorescent floss tied very small
Body Chocolate ostrich herl clipped down or swan herl dyed chocolate
Wing Grouse wing or tail fibres or speckled turkey
Hackle Dark furnace

Black Silverhorns

There are at least three species which the angler calls the Black Silverhorn: they are *Mystacides azurea*, *M. nigra*, and *Athripsodes aterrimus*. For angling purposes they can be treated as one and the same. The name Black Silverhorn is aptly self descriptive, for the insect is black with very long, silver banded antennae. Like the Grouse Wing the Black Silverhorns fly in very large numbers over the water surface during mating flight. This black cloud of sedges is very often totally ignored by the trout.

Larva

M. azurea: Case up to 15mm long comprising both vegetable matter and small stones, it often has one or two twigs running the length of the case.

M. nigra: Case up to 15mm long constructed out of sand grains and short dark vegetable debris, it is slightly curved.

A. aterrimus: Case up to 18mm long made entirely of sand grains, it is curved and tapers steeply towards the end, a very similar case to that of *M. longicornis*.

Adult Emergence *M. azurea:* end of May to August
M. nigra: mid June to mid September
A. aterrimus: June and July.

Artificials I would repeat that I have not seen a great deal of interest shown by the trout in these insects when they are on the wing, but occasionally a trout will rise and feed in the vicinity of the swarm. I believe the fish to be feeding on a spent sedge, or at least on one that has accidentally found itself on the water surface. I could be wrong, of course, and the trout could well be feeding on a totally different insect, but the following spent pattern may well be worth a try.

Spent Black Silverhorn (Price)

Hook	12 or L/S 14
Silk	Black
Body	Black polypropylene
Wing	Black raffene cut to shape and tied in to form a delta shape; the raffene is treated with fixative
Hackle	None
Antennae	Black and white teal fibres
Head	Two dots of red varnish either side to simulate the red eyes of the natural

Allow the fly to rest in the surface film and to move with the natural water or wind movement.

The other pattern was first published by John Henderson in the *Journal of the Fly Fishers' Club*; subsequently it appeared in John Veniard's *Reservoir and Lake Flies.*

Raffene
Artificial raffia, with more uses than we generally give it credit for, raffene has been mainly used as the material for creating the backs of such flies as the Polystickle. It also makes a very good wing case for nymphal patterns, however. Properly treated, the light coloured raffenes make an acceptable winging medium for some flies. The Americans have a similar product which they call Swish Straw, and some of their latest dry fly patterns use it.

Black Silverhorn (J. Henderson)

Hook	14
Silk	Gunmetal grey
Body	Black polymer or magpie tail
Rib	Pale green thread
Hackle	Palmered black cock trimmed down, black hen at the head

THE LARGE RED SEDGE (*Phryganea grandis, Phryganea striata*)

As you can see there are two contenders for the title of the Large Red Sedge. Both flies are big, brown coloured sedges with very little obvious differences as far as flydressing and fishing are concerned, and so they can be classed as one and the same. This fly is the 'Murragh' of Ireland, a very popular fly for fishing the large limestone lakes of that country. The Red Sedges are the largest of the British Trichoptera and can measure up to 27mm (wing length). The two species can often be identified by the dark blackish stripe that runs horizontally on the anterior wings; on *P. grandis* it is unbroken, on *P. striata* it is a broken line. Full identification can only be carried out microscopically, examining the configuration of the genitalia. This is usually beyond the scope and inclination of most anglers.

Larvae
P. grandis Case making: Case made from uniform fragments of leaves, spiralling in a taper to the distal end. Length between 30-50mm and approx. 8mm wide.
P. striata Case making: Size and construction of case similar to that of *P. grandis*, but a more parallel case.

Adult Emergence *P. grandis:* June to the beginning of August
P. striata: Mid May to end of June.

In all my insect hunting forays to various waters, I have yet to catch a specimen of either of these sedges to photograph. Like most people I cannot always arrange my holidays to coincide with emergent hatches of specific insects, but I'll catch one one day. (The brick and concrete wilderness of Sidcup is not the best hunting ground for sedges.)

Artificials An aquatic insect over an inch in length cannot possibly have been ignored by trout, anglers or flydressers, nor has it, for there has been a veritable regiment of artificials (both ancient and modern) tied to imitate the Large Red Sedge. The following are but a few.

Fixative
For some time the American dresser has used a product called Tuffilm to strengthen delicate and awkward winging materials. This product is readily available in this country in every art shop under the name of Fixative, and is used by artists to protect charcoal and pastel drawings; it also does a good job of preserving the appearance of dry fly wings. I use it quite often to coat raffene wings on dry flies in order to prevent them getting all wet and soggy, as is their wont.

Large Red Sedge
Flat wing style (Price)

Hook	L/S 10-8 or up eyed Mayfly hook of same size
Body	Grey brown polypropylene or fur
Rib	Gold oval
Body hackle	Natural dark red
Wing	Red hen body feather lacquered, tied flat
Hackle	Two natural red cock hackles, stalks left untrimmed to represent antennae

Red Sedge (R. Walker)

Hook	L/S 8-10 round bend
Body	Chestnut ostrich herl, clipped down, tipped with arc chrome dfm wool
Wing	A bunch of natural red cock hackle fibres, clipped square with the bend of the hook
Hackle	Two natural red cock hackles, the stiffer the better

Richard Walker's Red Sedge (above) can be tied in a variety of sizes to represent many of the brown coloured sedge flies that abound. In the larger sizes it can imitate the fly we are concerned with now.

T. Donald Overfield had the daunting and difficult task of bringing up to date *A Dictionary of Trout Flies* by Courtney Williams (A. & C. Black), by making a selection of modern flies for the fifth edition of this popular book. This task he fulfilled admirably, and amongst the new flies he chose is a Large Red Sedge, the creation of G. F. G. Rivaz of Hungerford, Berkshire, for fishing the large Irish limestone lakes.

Large Red Sedge
(G. F. G. Rivaz)

Hook	Not given, but one presumes a large hook
Body	Pheasant tail fibres
Rib	Fine gold wire
Body hackle	Furnace cock
Wing	Speckled turkey wing extending beyond the hook bend
Head hackle	Red cock, about eight to ten turns

The last pattern I give is the traditional Murragh of Ireland, for it is in that part of the British Isles that the Large Red Sedge has achieved greatest importance as a fishing fly.

The Murragh
(Large Red Sedge)

Hook	8-6
Body	Dark grey/black or black/claret mohair, or seal's fur
Wing	Dark brown speckled hen
Hackle	Two dark red rock hackles tied in front of wing

THE BROWN SEDGE (*Anabolia nervosa*)

This is an autumnal sedge, and as its name suggests it is of an overall dark brown coloration. Though large quantities hatch out from the end of August through to November, I have caught and photographed the insect as early as the first week in July. An interesting feature of this sedge is its great disparity of size. It can vary from about 10mm to 16mm. One of the visual identification features are the two small pale patches on the wing, one near the edge of the wing about half way along, the other patch approximately in the middle of the wing itself.

Larva A. nervosa builds a conical shaped case mainly out of grains of sand, to which it cements long pieces of twig. It is thought that these twigs prevent the larva being taken by predators. The length of the case including the twigs can be up to 60mm, the actual sand grain portion approximately 25mm.

Adult Emergence End of August to the beginning of October, although an early emergence can occur in July.

Brown Sedge (D. Jacques)

Hook	The creator recommends hook 14; for larger naturals step up in size to about a size 10
Silk	Chocolate brown
Body	Chocolate brown wool, silk, herl or dubbing
Rib	Fine gold twist
Body hackle	Chocolate cock
Wing	Game cock wing fibres rolled, sloping to back of the bend
Head hackle	Cock hackle dyed chocolate brown

Brown Sedge (Price)

Hook	10
Body	Mixed orange and brown seal's fur or polypropylene dubbing
Rib	Yellow terylene
Body hackle	Bright ginger
Wing	Dark brown hackle fibres
Hackle	Ginger cock

The first of the alternative patterns (above) is the creation of David Jacques; like his Pale Sedge given earlier his Brown Sedge is an artificial tied to represent a number of sedges, whose coloration in this instance is an overall brown.

Courtney Williams gives a pattern for a Little Brown Sedge, although the inventor of the fly was unknown to the author. Like the preceding pattern by David Jacques this fly does not imitate a specific natural insect, but is more of a broad spectrum fly imitating a wide number of sombre brown coloured sedges which are in evidence right through the fishing season. The name given for this pattern is the Little Brown Sedge, but as we are using hooks up to size 10 to imitate *A. nervosa* the description 'little' hardly fits the bill. For the sake of accuracy, however, I will stick with Courtney Williams.

Little Brown Sedge

Hook	14-10
Body	Orange tying silk with a mixture of brown and fawn wool
Rib	Fine gold wire
Hackle	Palmered from shoulder to tail
Wing	Red hen quill

BROWN SILVERHORN (*Athripsodes cinereus*)

This is a sedge of high summer; large numbers can be seen flying in clouds above the water, and they are found over both still and running water. John Goddard's researches have shown that the adult sedge is rarely found in trout autopsies but that the pupa is taken readily by the fish. This sedge is about the same size as the Grouse Wing Sedge, but without the distinct banding on the wing that the Grouse Wing has. The coloration of this sedge is, as its name intimates, brown, but the wing markings are very variable. Some specimens are an overall brown whilst others have small patches of yellow. The antennae, like those of the Grouse Wing and Black Silverhorns, are extremely long.

Larva Case made from grains of sand, curved and tapered.

Adult Emergence June to the end of August.
There are other species of sedge from the same family as *A. cinereus* that are very similar in appearance and habits; *A. aterrimus*, *A. commutatus* and *A. bilineatus*, to name but three, could all be considered as Brown Silverhorns from an angling point of view.

Dressings Like the Black Silverhorns it would appear that the trout do not often take the adult, but I do not think that the trout would turn down a meal if one of these flies should find itself trapped on the surface. The dressing I am giving is for just such an occasion.

Brown Silverhorn Spent (Price)
Hook 14
Body Brown polypropylene or fly body fur
Wing Light brown raffene treated with fixative
Antennae Brown mallard
Hackle None
Soak this fly well in floatant, and allow it to fish for itself

As there is some doubt about the value of adult sedge imitations to represent the Brown Silverhorn, I shall only give one alternate pattern in this instance. It is one of John Henderson's creations, given in *Reservoir and Lake Flies* by John Veniard.

Brown Silverhorn (J. Henderson)
Hook 14
Silk Brown
Body Green polymer mixed with hare's ear
Rib Gunmetal grey
Body hackle Rusty dun trimmed from $\frac{1}{4}$in. to $\frac{1}{16}$in.
Head hackle Woodcock neck feather (Henderson gave landrail as alternative, but this is now a protected bird)

THE CAPERER (*Halesus radiatus, Halesus digitatus*)

Once more we are back with the larger sedges, not quite as large as the Great Red Sedge, but quite large for all that, reaching around 20-22mm. There are two closely related insects (the species named above) both with the common name Caperer. This sedge probably received its name because of its method of skittering across the surface after hatching. These sedges hatch out in open water, and instead of taking to the wing they race across the water surface to seek the shelter of the banks. It is at this time they can well interest the trout. The overall wing colour of this insect is yellowy brown, while the body can vary from a brown with an orange tinge to it to a brown with a hint of green. A noticeable dark brown line is apparent on the top of the wing when the insect is at rest, and this can be used as a rough point of identification for *H. radiatus.* The photograph in the colour plates is of its cousin *H. digitatus.*

Larva

H. radiatus: Case made from small pieces of leaves and stems of water plants, often with pieces of bark; along the length of the case one finds one to three twigs; the overall length can reach 38mm including twigs; the case measures about 33mm without.

H. digitatus: Case made from a mixture of materials, including sand grains, small stones, leaves and other vegetable matter; like the other species a twig or two is often included; the case is a little smaller than the other species, tapering slightly, with a length of up to 27mm.

Adult Emergence

H. radiatus: September to end of October

H. digitatus: September to November. The specimen shown in the colour plates hatched out of my aquarium two days before December 1st.

Dressings As far as specific dressings for the Caperer are concerned, strangely enough they are very few and far between. The one that is always given is the dressing by William Lunn, sometime keeper on the Test. He bracketed his Caperer with another misnamed sedge, the Welshman's Button. The resultant fly had a chocolate brown body with a centre band of yellow and a mixed black and brown hackle. This fly has stood the test of time and continues to catch fish, but an imitation of the Caperer it most certainly is not. The trout take it, I think, because of its overall edible appearance; with its yellow band it looks more like a hoverfly.

I had better explain why I think the Welshman's Button is a misnomer. For a start a sedge bears no relation to a button of any description, let alone a Welsh one. The one creature that fills the bill, button wise, is of course the old favourite the June bug or Garden chafer. This is perhaps better known to anglers as the Coch y bonddu, a beetle.

For artificials of the Caperer try any large brown sedge; the Wake Sedges described earlier fill the bill admirably. Fish them in the manner suggested so that they cause a wake on the surface of the water; this is how the natural insect behaves.

For an imitation to represent both the Caperer and other sedges such as the Large Cinnamon the following should suffice.

Caperer (Price)

Hook	L/S 8-10
Silk	Orange
Body	Orange seal or poly dubbing mixed with hare's ear
Rib	Orange silk
Body hackle	Cree or light ginger
Wing	Cree fibres clipped to shape
Hackle	Cree (two)

THE WELSHMAN'S BUTTON (*Sericostoma personatum*)

Reputedly it was the great dry fly exponent of the chalkstream, F. M. Halford, that named this sedge the Welshman's Button. Why we shall never know, for as explained earlier the Welshman's Button can only be the Coch y bonddu or a similar member of the Coleoptera.

This sedge is most common on rivers, but I have taken specimens from the surface of a local lake. It is possible that these flies had originally hatched in the river that feeds the lake, but as nearly all still waters are fed by river, stream, or brook then river-living species can often be found on the still waters that they create. This sedge is quite a dark insect, both in the wing and body, and is generally a deep shade of brown. They have robust antennae which are only about the size of the wing, occasional specimens are a lighter shade of brown, and like many species of sedges a variance in the body coloration can occur.

Larva
Case made entirely of sand grains, tapered and curved and about 16mm in length.

Adult Emergence From the beginning of June to the end of August, often hatching in daylight hours.

Dressings

Welshman's Button (Price)

Hook	12
Silk	Black
Body	Chocolate coloured turkey herl
Rib	Gold oval tinsel
Wing	Flat wing style made of a pheasant cock body feather
Hackle	Natural red game

Dark Sedge (T. Thomas)

Hook	4-14
Body	Black wool or chenille
Body hackle	Palmered black cock
Wing	Black deer hair tied by the tips and clipped square with the bend
Hackle	Black cock

Welshman's Button (J. Henderson)

Hook	10-12
Silk	Chestnut brown
Body	Cock pheasant tail fibres
Rib	Gold tinsel
Body hackle	Trimmed rusty dun cock tapering from $\frac{5}{8}$in. to $\frac{1}{4}$in.
Hackle	Copper cock pheasant breast feather, or Rhode Island Red cock

As a first alternative pattern I have chosen another broad spectrum sedge by Terry Thomas. It is his Dark Sedge, a fly that can be used to imitate any dark natural sedge. The final dressing for this sedge is one of the John Henderson creations given in *Reservoir and Lake Flies* by John Veniard.

Other Sedges

I have mentioned but a few of the 197 or so species of sedges we have in the British Isles, so it goes without saying that many other sedges must form part of the trout's diet. Many of the sedges I have not mentioned are either too small to be of interest to the angler or so local in distribution as to be of little consequence.

A couple of exceptions to what I have just said would be the Little Red Sedge, *Tinodes waeneri*, a fairly well distributed insect that has received the attentions of numerous flydressers. The following pattern is one of David Jacques' and called simply the Little Red Sedge.

Little Red Sedge (D. Jacques)

Hook	12-14
Silk	Orange
Body	Hare's ear fur
Rib	Gold twist
Body hackle	Stiff red cock hackle
Wing	Rolled landrail substitute
Hackle	Deep red cock hackle

Another sedge which I have seen in large numbers hatching on Darwell reservoir in Sussex could be classed as a small red sedge. In this case the species was *Limnephilus vittatus*. David Jacques' Little Red Sedge would have been quite sufficient to imitate this fly; on the occasion when I first noted this sedge the only flies that would work were small sedge pupae and Amber Nymphs.

A largish sedge that inhabits fast flowing water is the sedge *Odentocerum albicorne*, known as the Grey or Silver Sedge; a much smaller sedge that inhabits still water, *Lepidostoma hirtum*, is also referred to as the Silver Sedge, and can occur in large enough numbers to interest the trout. The standard pattern for this fly is as follows.

Silver Sedge

Hook	14
Body	White floss
Rib	Silver wire and ginger palmered hackle
Wing	Landrail substitute over body
Hackle	Light ginger

The Mayflies

ORDER: Ephemeroptera

No matter what species of the Ephemeroptera we see, I do not think I know a single angler who does not admire the perfect beauty of these jewel-winged insects. Those who are not moved by these creatures as they take to wing, then truly they are not really anglers at all and certainly do not deserve to witness the glory and miracle of the hatch. Sadly, on still water their importance does not always match their perfection; there are of course exceptions, the Mayfly proper on the large Irish Loughs and perhaps the small white curse of all fly fishermen, the Caenis, whose spinners confetti down onto the surface to be taken greedily, much to the frustration of anglers. No, on still water the trout have more abundant fare. Creatures such as the Chironomid midge, the minute daphnia, sedges, and even the slow moving snails, are the usual, basic diet, judging from the stomach contents of caught fish.

There are 47 species of Ephemeroptera on the British list, but only the following few interest the stillwater angler. Each will be dealt with separately in due course.

The Mayfly: Ephemera species, usually *Ephemera danica*
The Sepia Dun: *Leptophlebia marginata*
The Pond Olive: *Cloeon dipterum*
The Lake Olive: *Cloeon simile*
The Caenis (Angler's Curse): one of five British species
The Claret Dun: *Leptophlebia vespertina*
The Summer Dun: Siphlonurus species, usually *Siphlonurus lacustris.* (But on waters in Sussex where I have fished. *S. armatus* hatches in fairly large numbers.)

Added to the above, some species that are usually found in flowing water can also occur on some still waters; the ubiquitous Blue Winged Olive, *Ephemerella ignita*, is one such fly. Another that occurs in small numbers on a local lake where I am privileged to fish is the Yellow May Dun, *Heptagenia sulphurea.* This fly is sometimes called the Yellow Hawk, and I believe the flies I have seen may have originated from the river that flows into the lake.

General Identification

All the Mayflies are slim bodied flies possessing ultra-long tails, three in some species, two in others. The wings are held upright over the body when at rest; most species have four wings, the front pair being the larger, whilst the hindwings are very much smaller. In some species they are non-existent and in others they are merely spur-like appendages (hence the name 'Spurwing'). Fossil evidence shows that prehistoric members of the Ephemeroptera had four wings of equal size.

Unlike the sedges the mayflies do not possess long antennae. What of size? Well, members of this family can vary from very small flies of about 5mm to spectacularly large insects like *E. danica* that measures about 25mm excluding the setae (tails).

Life Cycle

Unlike other insects such as the Trichoptera (sedge flies), Lepidoptera (moths and butterflies) and Coleoptera (beetles), the members of the order Ephemeroptera have an incomplete yet unique metamorphosis. Unlike the other groups mentioned they have no pupal stage. The life cycle is as follows:

Egg → Larva → Subimago → Imago

In angling terms we know these stages as:

Egg → Nymph → Dun → Spinner

Eggs are laid in the water in a number of different ways, depending on the species. Some females fly over the water surface, dipping the tips of their abdomens in order to wash off individual eggs. In other species the female releases an egg ball as she flies over the water surface, or she dips her abdomen beneath the water to wash off the ball of eggs. A third method adopted by some species is to submerge themselves completely under the water to lay the eggs there. As soon as the eggs reach the bottom they adhere to the stones or gravel; there they remain for a week or so, before hatching into the larval or nymph stage.

Larva The larval stage of this insect is always referred to as the nymph. The larger of the Mayflies can spend up to two years as a nymph. Other species remain in this stage for one year, whilst some others can have two or three broods in 12 months. The nymphs of some species, and *E. danica* is such an insect, are adapted to burrowing, their front legs being powerfully built

1.1 Bloodworm larvae

1.2 Chironomid pupae

1.3 Chironomid midge (female)

1.4 Chironomid midge (male)

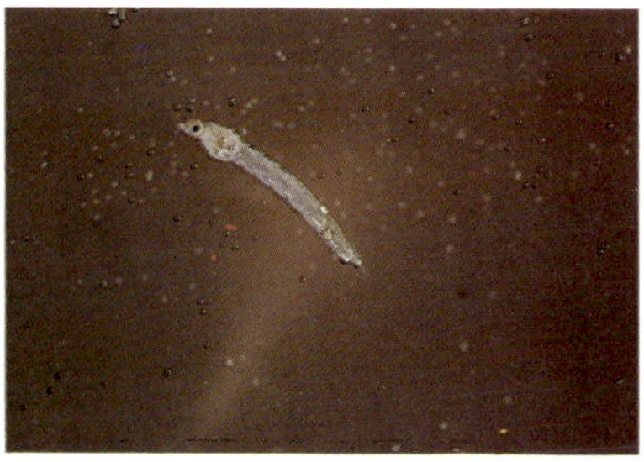

1.5 Phantom larva

1.6 Phantom pupa

1.7 Phantom midge (female)

1.8 Phantom midge (male)

1.9 Mayfly nymph

1.10 Mayfly dun

1.11 Mayfly spinner

1.12 Pond olive nymph

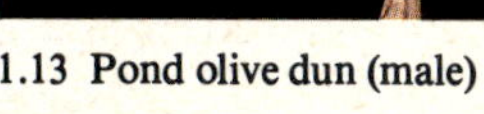

1.13 Pond olive dun (male)

1.14 Pond olive spinner (female)

1.15 Sepia nymph

1.16 Sepia dun

1.17 Sepia spinner

1.18 Summer dun nymph

1.19 Summer dun

1.20 Summer dun spinner

1.21 B.W.O. spinner (male and female)

1.22 Spent caenis

1.23 Yellow May dun

1.24 Caddis in case

1.25 Caddis larva out of case

1.26 Caddis in case

1.27 Swimming caddis larva

1.28 Sedge pupa

1.29 Spent pupal case

1.30 Grouse wing sedge

1.31 Cinnamon sedge

1.32 Red sedge

1.33 Mottled sedge

1.34 Large cinnamon sedge

1.35 Brown silverhorn

1.36 Black silverhorn

1.37 Ronalds' sand fly sedge

1.38 Caperer sedge

1.39 Brown sedge

1.40 Welshman's button sedge

1.41 Freshwater shrimp

1.42 Freshwater louse

1.43 Daphnia

1.44 Water boatman

1.45 Corixa sp.

1.46 Marabou Bloodworm

1.47 Footballer

1.48 Mayfly Nymph

1.49 Straddlebug Mayfly

1.50 Yorkshire Mayfly

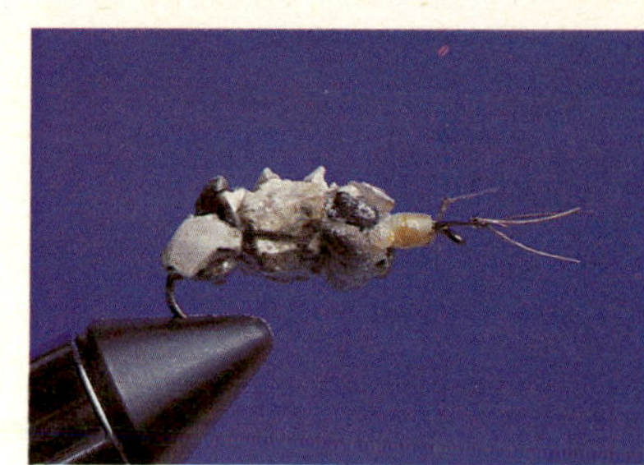

1.51 Pebble Caddis Larva

1.52 Latex Caddis Larva. Sand Caddis Larva

1.53 Goddard's Sedge Pupae

1.54 Goddard's Sedge Pupa. Sweetie's Sedge

1.55 Wake Flies

1.56 G & H Sedges

1.57 Deschutes Caddis

1.58 Large Red Sedge (flat wing)

1.59 Douglas Water Louse (*bottom*) Olive Shrimp; and (*right*) Gold Shrimp

1.60 Orange Nymph

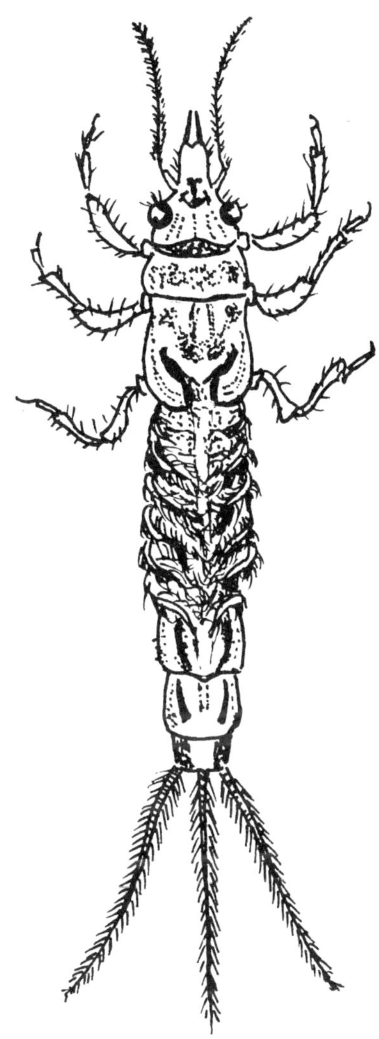

Ephemera nymph (Mayfly)

for this task. Other nymphs are adapted for crawling around the bottom. Some river species have crablike legs for clinging onto stones in the strong current and yet others swim quite freely from weed bed to weed bed. Some species are extremely agile, whilst others appear slow and laboured in their movements. Sometimes these nymphs are described by their apparent habits, i.e. stone clingers, burrowers, agile darters, weed clingers, etc., but all nymphs can, in given circumstances, be all of these things. The nymph of *E. danica* taken out of its burrow can be quite agile and swim very well. A stone clinger can be just as happy crawling along the bottom.

The nymphs feed on a wide variety of plant matter, algae and other allied life forms, and some are believed to be partially carnivorous. In order to grow and get larger the nymphs undergo a series of moults; each time the rudimentary wings get bigger. When it is time to hatch out the nymph rises to the surface by swimming, or by crawling up reeds, stems and the like. Its larval skin splits and the subimago or dun emerges.

The Dun This is the stage that can be equivalent to the pupa in other insect forms, even though the Mayfly does not stay long as a dun. As soon as its wings are dry the dun takes to wing and seeks the shelter of the bushes and trees. The slower ones that stay on the water too long often fall victim to the ever waiting trout, and there are many that run the gauntlet of feeding fish and take to the wing only to be preyed on by the many birds that consider these slow flying insects to be a good and easy mouthful. Those flies that have reached shelter rest up in their green haven. The final moult in this insect's life takes place and the full adult emerges.

The Spinner If the dun is a beautiful fly, then the spinner, like some pre-midnight Cinderella, is even more beautiful. The duller coloration of the subimago gives way to a bright shining star of a fly with wings fashioned as though from crystal. This is the mating stage, and the males and females of the species are adorned for such an occasion. Incidentally, the eyes of the males are always very much bigger than the females'. I have never found out why—'all the better to see you with', perhaps. The males of the species swarm usually, but not always, in the evening. I have seen them at midday dancing in the sunbeams, with the delightful, undulating flight typical of this insect. They seem to soar right up into the sky like larks, then free fall, only to rise once again. These swarming males are joined by the females, and mating takes place high on the wing and lasts for but a minute or so. It is usually all over before the descent. The female returns to the water to lay the eggs. This task done she slowly dies, lying on the water with wings spread out, her life force spent. She has become the spent spinner, and once more receives the attention of hungry trout.

THE MAYFLY (*Ephemera danica*)

I have not started with this insect because I think it the most important member of the family for the stillwater angler. No, I have chosen it because of its spectacular size. It truly is a large, impressive insect.

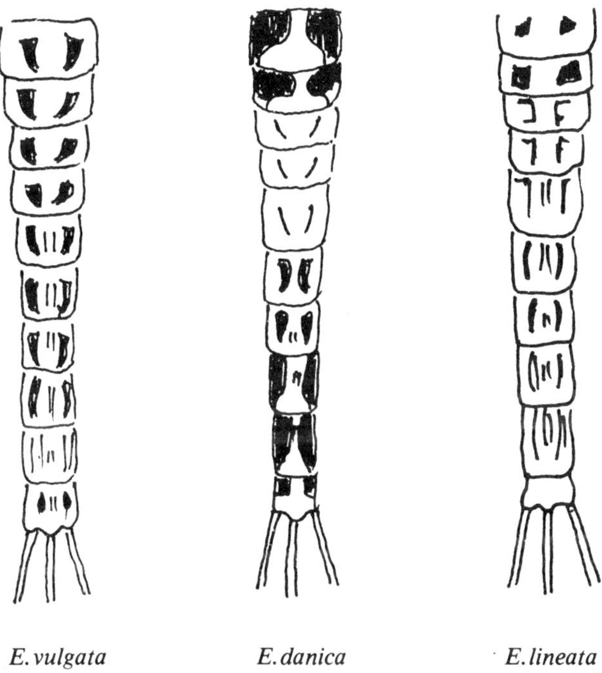

E. vulgata *E. danica* *E. lineata*

In the introduction to the chapter on Ephemeroptera it was stated that there were in fact three species:

Ephemera danica	By far the most prolific, found in most parts of the country
Ephemera vulgata	Less common than *E. danica*, prefers slower muddier rivers, as in the eastern counties
Ephemera lineata	Extremely rare and therefore of little importance for the angler

By examination of the ventral segments of the abdomen, it will be seen that the three species have different markings.

A problem sometimes occurs with tyros in the gentle art; they find difficulty in understanding some of the common names given to some of the stages of this species. The following list may well be of help.

Grey drake	Spinner (imago) (female)	*E. danica*
Green drake	Dun (subimago) (male)	*E. danica*
Black drake	Spinner (imago) (male)	*E. danica*
Spent gnat	The dead female spinner after egg laying	*E. danica*
Spent spinner	As spent gnat	*E. danica*

General Identification (*E. danica*)

Nymph The full grown nymph (the larval form, remember) measures 20-24mm long. The colour is best described as a dirty cream with dark brown markings. These are in the form of triangles on the 7th, 8th and 9th abdominal segments. The legs are developed for digging and are, therefore, short and powerful. The mouthparts are beak-like in their formation.

Their habitat is usually clean, gravel bottomed rivers with a fairly fast, well oxygenated flow. They are found also in clean lowland (usually chalk or limestone) lakes.

The Dun The body is creamy yellow with a tinge of olive, with distinct brown markings on the abdominal segments. Large forewings are heavily veined with olive shading. Hind wings are much smaller, but unlike those of smaller members of this family they are easily seen. Both sets of wings are held upright. It has three long tails.

The Spinner The body is ivory white, its markings dark brown, almost black, and there are no markings on the first four abdominal segments. The wings become more transparent in the imago stage, with black veining and patches coloured a dark olive brown. The tails increase in length to over 1½ times body length.

Adult Emergence End of May to second week of June. (On occasions it has been known to hatch out at the end of April. At other times it is in evidence as late as July. The odd specimen has been noted much later in the year.)

Nymphal Imitations

Flydressing books abound with nymphal dressings of the Mayfly. They vary in colour from white to brown, and some I have seen are almost yellow. They are all called Mayfly nymphs, many catch fish, but very few imitate the larval Mayfly. The first pattern is one of my own (it's one of the perks when compiling a book of this sort, you can put your own flies first). This fly has caught fish for me, both browns and rainbows, on rivers and also on still water. It is easy to construct and, furthermore, it approximates the correct coloration of the natural nymph.

Mayfly Nymph (Price)

Hook	L/S 10 (can be weighted if desired)
Silk	Brown
Tail	Cock pheasant tail fibres (three)
Body	Cream nylon wool
Rib	Brown ostrich herl with two turns of ostrich herl just before the thorax
Wing case	Cock pheasant tail fibres
Thorax	Cream wool
Legs	Some of the cock pheasant fibres from the wing case turned down beneath the hook

The next pattern is one of those traditional flies that bears no resemblance to the natural but catches fish, so it deserves to be included for that reason.

Mayfly Nymph

Hook	L/S 10
Tail	Three cock pheasant tail fibres
Body	Three turns of dirty yellow seal's fur, the rest brown seal's fur
Rib	Yellow silk
Thorax	Brown seal's fur
Wing case	Hen pheasant tail feather
Hackle	Hen pheasant neck feather

Richard Walker, in his popular column in *Trout and Salmon*, gave a very good imitation for the Mayfly nymph. This fly he improved on, and the resultant imitation was published in his book *Fly Dressing Innovations* (Ernest Benn).

Richard Walker, *Fly Dressing Innovations*, Ernest Benn.

Mayfly Nymph (R. Walker)

Hook	L/S 8
Silk	Dark brown
Tail	Five pheasant tail fibres
Body	Five strands of cream ostrich herl and the pheasant tail butts
Wing case	Pheasant tail fibres
Legs	Pheasant tail fibres

The tying procedure for this fly is as follows:

Tie in the pheasant tail fibres for the tail; follow this by tying in the five strands of ostrich herl. Wind two turns of ostrich herl and tie in; now tie one turn of the twisted butts of pheasant tail fibres and secure; now follow that turn with two further turns of the cream ostrich, and then by one turn of pheasant tail; cut off the remainder of the latter, and continue to wind the ostrich herl to about three fifths of the way up towards the eye. Now tie in a bunch of pheasant tail fibres for the wing case—the ends of the fibres will later form the legs. Wind more ostrich herls to form the thorax, take the pheasant tail fibres over this thorax, and finish off in the usual way.

The next pattern is a hybrid, based on the pattern I gave earlier, and on the style of articulated fly devised by Swisher and Richards, authors of *Selective Trout*. This style of 'wiggle nymph' was further popularised by Dave Whitlock. Step by step instructions will show clearly the procedures used in constructing this realistic style of nymph, and the process is also shown in an illustration sequence in Book 2.

The Wiggle Mayfly Nymph

Hooks	Tail hook (Barb removed to bend) L/S 10-12; front hook, down eyed normal 10 (weighted). If possible use a ball eyed hook at the rear
Silk	Brown
Body	Rear hook only, cream nylon wool or polypropylene
Rib	Brown ostrich herl
Front Hook	Two turns of ostrich to simulate breathing filaments
Thorax	Cream wool, etc.
Wing case	Pheasant tail fibres, the ends used as legs

Tying procedure

1. Rear hook in vice, wind silk to the end, tie in three pheasant tail fibres and secure; 2. Tie in wool body material and one strand brown ostrich; 3. Wind body to the eye; 4. Wind on rib. Finish off. Cut a small length of trace wire; 5. Feed it through the eye of the hook and by means of pliers twist; 6. Remove the hook from vice and with the pliers snip off the hook at the bend; 7. Place the front hook in the vice and wind on the tying silk; 8. Place twisted wire on the top and run tying silk up and down to secure. Varnish these turns of silk; 9. Tie in strand of ostrich herl and give two turns; 10. Tie in pheasant tail fibres and cream wool; 11. Wind on cream wool; 12. Take pheasant tail fibres over back and tie off; 13. Bend back pheasant tail fibre ends to form legs; 14. Finish with whip finish, varnish, and the fly is complete.

Fishing the Nymph

The Mayfly Nymph, as we know, is a creature that burrows into the mud. For most of the time, therefore, it does not receive the attention of the trout, for it is not swimming around free except just before hatching. So it goes without saying that one should endeavour to fish the nymph when the first few winged duns are seen. Quite often prior to hatching the nymph will make an abortive sortie to the surface, perhaps to practise for the great event of its life. I like to fish the Mayfly Nymph when I notice the first of the Mayflies on our lake; it indicates that there must be one or two nymphs on the move, and if the dun is ignored by the trout as something strange and alien, and untried as food, they do not often ignore a nymph. Use as long a leader as possible in conjunction with a floating line, allow it to sink slowly (often a take can occur on the drop), and retrieve the fly in slow, long, steady pulls, as though the fly was a real insect swimming up to the surface.

Artificials for the Mayfly Dun

Early dressings for this stage of the insect's life usually made use of the body feathers of the mallard drake, hence the predominance of the word 'Drake' in many of the artificials seen in the flydressing books. Fanwing Mayflies using such feathers do catch fish, but I am not alone in feeling that they tend to catch more fishermen than trout. All are attractive looking flies begging to be bought, even though many will end up in the hat rather than in the scissors of a hungry trout's mouth.

The following patterns are all dry flies; I will give a few wet fly patterns at the end of this piece on the Mayfly.

One could fill a book with dressings for the Mayfly. Just look at some of the fishing catalogues of bygone days to see the prolificacy of different Drakes and Spinners that were depicted on their pages. Two non-winged imitations that I like are the French Partridge Mayfly and the Straddlebug Mayfly. Both are killing flies on their day.

Mayfly Dun (Price)
Hook Up eye Mayfly hook, 12-10
Silk Black
Tail Three moose mane strands
Body Dirty white polypropylene dubbing
Body hackle Olive
Head hackle Mixed badger and olive cock hackles

French Partridge Mayfly
Hook L/S May 12
Silk Black or brown
Tail Three red cock fibres
Body Cream floss (alternative pattern uses natural raffia)
Rib Cross rib of gold tinsel and red (crimson) tying silk
Hackle Inner, natural red cock; outer, French partridge

Another variation has palmered olive cock down the body.

The Straddlebug Mayfly has the colour orange within its makeup. Why the colour orange weaves a magic spell on trout I do not know, but many stillwater flies and lures, and even delicate dry flies for chalkstreams, all with the colour orange, seem to be excellent killing patterns. The Whisky Fly, Jersey Herd, Dunkeld, Orange Quill, etc., etc., are all flies that are orange in some way. The Straddlebug is no exception—it too is a very good trout-taking fly.

Straddlebug Mayfly
Hook L/S May 12
Tail Three black cock hackle whisks
Body Raffia (natural)
Rib Brown silk
Hackle Inner, hot orange; outer, summer duck (or substitute)
Head Peacock herl

John Goddard has a pattern for the Mayfly which he has called the Nevamis—I bet it does sometimes! The dressing is alongside.

The next fly is also a modern pattern, devised by Peter Mackenzie-Philps whose effective sedge pupa pattern was given earlier. This Mayfly can be used to imitate either the dun or the spinner, and looks extremely realistic. Peter had not given his fly a name up to the time of writing this piece so perhaps he will forgive me for calling it the Yorkshire Mayfly.

Nevamis Mayfly (J. Goddard)
Hook L/S 8 Mayfly
Tail Three pheasant tail fibres
Body Cream seal's fur wound thickly
Rib Oval gold tinsel
Hackle Palmered honey cock, then clipped from $\frac{1}{4}$in. to $\frac{3}{8}$in., head to tail
Wing Pale blue dun hackle fibres tied in a V formation
Head hackle Short fibred Coch y bonddu or furnace hackle

Yorkshire Mayfly
(P. Mackenzie-Philps)
Hook Up eye dry fly 8 or Yorkshire flybody hook
Tail Three chocolate labrador dog hairs (a substitute will have to be found if you haven't such a dog lying around)
Body Thin strip of Ethafoam cut with the grain and wound around the shank. The body can be then marked with a black felt tipped pen
Wing Short tufts of firebrand white fluorescent floss tied in with a figure 8 and cut to approx. $\frac{1}{4}$in. This simulates the roots of the natural wing, the rest of the wing being left to the trout's imagination
Hackle Large badger cock trimmed underneath. This allows the fly to fall the right way up all the time, and also to float on the surface in a very natural manner

For those flydressers who would like to tie other patterns I recommend any of the John Veniard books on flydressing. His *Fly Dressers' Guide* and *Further Guide to Fly Dressing* are both volumes which abound with dressings for the fly which is the embodiment of fly fishing.

John Veniard, *Fly Dresser's Guide*, A. & C. Black.

John Veniard, *Further Guide to Fly Dressing*, A. & C. Black.

Artificials for the Spinner

Just like the dun, the spent spinner is represented in flydressing books by a host of patterns. You can truly say that the Mayfly has caught the imagination of flydressers more than any other fly.

Spent Spinner (Price)
Hook L/S 12
Tail Three cock pheasant tail fibres or moose mane
Body White polypropylene
Rib Black silk
Wing White deer hair tied flat on the top

It will be noted that the above pattern has no hackle. Instead the fly is supported by the buoyant deer hair, and it rests in the surface film in a most realistic fashion.

Spent Female Mayfly (R. Walker)
Hook Fine wire L/S 10
Silk Dark brown
Body Polyethylene foam
Rib Brown silk
Wing Blue dun hackle points tied flat
Hackle Furnace cock

Spent Mayfly (J. Veniard)
Hook 10-12 L/S
Tail Three cock pheasant tail fibres
Body White floss silk or raffia
Wing Blue dun hackle points
Hackle Badger

The second pattern is from the Richard Walker book *Fly Dressing Innovations*. Like the Yorkshire Mayfly given earlier it uses a modern plastic body medium.

The final Spent Mayfly dressing I give is one of the John Veniard traditional patterns, a satisfying pattern to dress and an effective fly to fish with.

Wet Mayfly Patterns

One of the traditional methods of fishing the Mayfly is as a wet fly. Presumably this method represents one of two natural events. Firstly, it may imitate a drowned Mayfly. Quite often the adult insect fails to emerge from its nymphal shuck; it can be caught up within the shuck, the insect finally dying. This stage has provided the American anglers with a host of what they term 'stillborn flies'. Secondly, it may seem to be a nymph

The Gosling
Hook 8-10
Silk Brown or olive
Tail Brown mallard fibres
Body Yellow seal's fur
Rib Oval gold
Hackle Inner, hot orange; outer, grey mallard

in the process of hatching, if the fly is fished just under or within the surface film.

The first of these wet patterns is the famous Irish fly known as the Gosling. (There is an American pattern by the same name which looks nothing like the Irish version.)

Another traditional wet pattern given by John Veniard in his *Fly Dressers' Guide* is as follows.

Hackled Wet Mayfly
Hook 8-10
Silk Yellow
Tail Cock pheasant tail feathers
Body Yellow dyed lamb's wool
Rib Oval gold
Hackle Buff coloured speckled feather from a hen pheasant breast, tied halfway down the body

Those, then, are the dressings for the Mayfly. They are but a few of the many to be found in the angler's library. All are extremely satisfying flies to tie, and fun to fish with.

Fishing the Mayfly

On the smaller still waters, such as the modern 'put and take' fisheries where a hatch of Mayfly is an annual occurrence—some of those in Hampshire are examples of such waters—then conventional fly fishing from the bank to rising, feeding fish is the usual method to adopt. Bear in mind one important point—concealment! Station yourself behind convenient reed or sedge clumps. If the water is completely open, you will do better if you kneel to cast. Watch the cruising trout and endeavour to place your dry fly, whether it be a dun or spinner imitation, a yard or so down its path, then hold your breath.

What has just been said for Mayfly fishing holds true for any dry fly under the same circumstances. (It does not matter whether the trout are preoccupied with Mayflies, Pond Olives or even a fall of ants, the same basic principles apply.)

Dapping Before passing on to other Ephemerids mention must be made, if only briefly, of a method of fly fishing which was the forerunner of our modern sport. In the early days of fly fishing, casting was unknown. The angler relied on getting his fly, which more often than not was attached to a fixed line (as in modern roach pole tactics, but without the rubber shock attachment), onto the water by means of wind and the length of his rod. This method was dapping pure and simple, and it is still an effective method of taking trout, especially those trout in large lakes and reservoirs. If you have a decent breeze then you can dap. In Ireland, on lakes such as Mask and Corrib, dapping with the Mayfly, and later in the season with Crane-flies, is a long established and popular method of catching fish.

For dapping one needs a long rod. Many anglers make their own using light coarse fishing rods. No fly line is used; instead a flat floss silk line the same length as the rod (anything from 14 to 16 ft) is attached to the line backing, and at the other end a tapered cast is also attached. At no time does the line touch the water. The wind bellies out the floss, allowing the fly to skitter tip-toe across the surface, and the rod is fished in a slow arc in front of the boat, covering as much water as possible ahead of the boat as it drifts. At no time is the rod put down; it must be held at all times, for a take can come when you least expect it. This method of fishing can be extremely tiring, for not only are you holding your rod at all times, your eyes must not stray from the dapping fly. It is a fishing method that requires complete and utter concentration. When the take comes most people strike too soon, so pause before striking, or at least until the line in front of you tightens. Then it is time to uplift the rod and tighten into the fish. Many anglers, when they get such a take, have pet sayings or phrases to aid their timing, such as 'God Save the Queen' or 'One two three strike'. They have equally familiar phrases if they miss the fish, most of them unprintable.

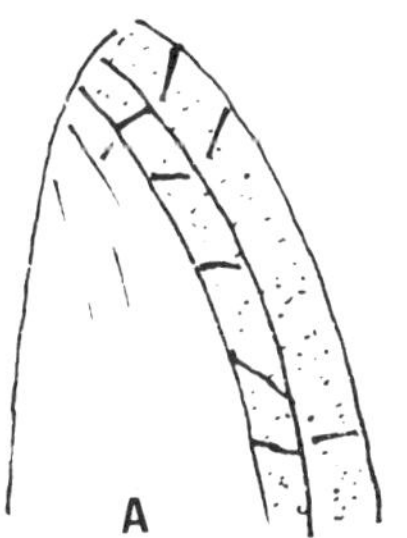

THE POND OLIVE (*Cloeon dipterum*)

After the largest of the Ephemeroptera we come to one of the smaller members of this order. If you live in Ireland then in all probability you will not have come across this insect, for it is not a creature of the Emerald Isle. Its main habitat is in the south-east of England, and in fact we have a very good hatch on our lake in Kent. It occurs elsewhere, but does tend to be confined in the eastern part of the country, except for an occurrence (along with its allied fly the Lake Olive) in the north-west. This fly favours smaller waters, such as small shallow lakes, ponds and the smaller of our reservoirs. D. E. Kimmins des-

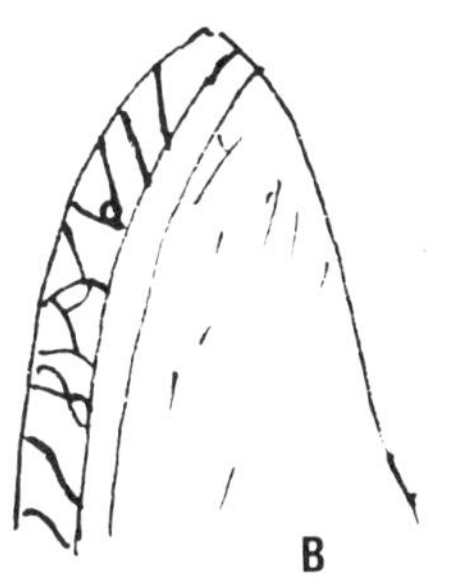

Leading edges of wings:
A—*C. dipterum*
B—*C. simile*

cribes, in *A Revised Key to the Adults of the British Species of Ephemeroptera* (Freshwater Biological Association No. 15), how this insect established itself in the static water tanks erected to supply water during the blitz in the 1939–45 war.

D. E. Kimmins, *A Revised Key to the Adults of the British Species of Ephemeroptera*, Freshwater Biological Association Publication No. 15.

Description and Guide to Identification

The Nymph The Pond Olive nymph is 7.5-10.5mm long. Like all nymphs of British Ephemeroptera it has three tails, which are held well apart and are fringed with fine hairs. Near the middle of the tails there is a distinctly darker band. The overall colour of the nymph varies with the habitat; it is usually an olive colour but can be brown in some cases. It is found in the weedy areas of the lake and is an excellent swimmer. It has seven sets of breathing gills running the length of the abdomen, and I believe the function of the seventh set is to redirect the water back over the others, presumably to obtain more oxygen.

The Dun The dun is about 8-9mm long. The only other insect this dun could possibly be confused with is its near relative the Lake Olive. The insect possesses two tails only, and they are ringed distinctly with dark brown. The Pond Olive possesses only one pair of wings, which in the dun stage are a dark grey. At rest the insect holds its wings slightly apart in a 'V' formation, unlike many other mayflies which hold them close together. The overall body coloration is an olive shade with occasional splashes of browny orange.

The Spinner We consider only the female spinner—the male is comparatively unimportant as a fishing fly. The angling name now given for this final stage in the insect's life is the Apricot Spinner, because—yes, you've guessed it—it is of an apricot colour. A distinguishing feature of the fly is the yellowish brown colour of the leading edge of its wings. I have seen one specimen of the spinner which was still an olive colour, with the brown wing edge distinctly green.

Adult Emergence End of May to July.

C. F. Walker in *Lake Flies and their Imitations* gives one nymph for both the Pond and Lake Olive. This fly he has termed the Cloeon Nymph.

Nymphal Imitations

Pond Olive Nymph (Price)
Hook 16-14
Silk Brown
Tail Dark olive cock hackle fibres
Body Olive brown fur
Thorax The same
Wing case Dark brown hen fibres

Cloeon Nymph (C. F. Walker)
Hook 14-16
Tail Fibres from any brown speckled feather
Abdomen Mixed brown and ginger seal's fur mixed, kept smooth
Gills A ribbing of pale brown condor herl
Rib Silver tinsel
Thorax and Wing pad Dark brown seal's fur
Hackle Medium honey dun
Again by using olive fur instead of the mixture described a lighter coloured nymph is obtained.

Pond Olive Nymph
(A.C.R. Howman)
Hook 12-13
Tail Clipped olive hackle
Body Green nylon
Legs and Wing case Green ostrich herl

The Dun On the lake where I fish I have yet to see a trout take the dun of the Pond Olive. Our fish seem preoccupied with feeding on chironomid pupae and appear to leave this delicate insect well alone. Just because I have not seen my local trout feed on this insect does not mean to say that, on other waters at other times, trout will not feed quite happily on this dun. Judging from the number of successful patterns there are for this fly they must do.

Bob Church, *Reservoir Trout Fishing*, Cassell.

Bob Church is a household name where boat fishing for reservoir trout is concerned. In his book *Reservoir Trout Fishing* (Cassell) he feels it an important enough fly to give a dressing for the dun stage of this insect.

Pond Olive Dun (Price)
Hook 14 or Yorkshire Fly body hook
Silk Olive
Tail Blue dun
Body Olive swan or goose herl
Rib Brown silk
Wing Blue dun hackle fibres upright
Hackle Olive

Pond Olive (Bob Church)
Hook 10-14
Tail Olive cock hackle
Body Two strands of pheasant tail fibres
Wing Light starling tied upright
Hackle Olive cock

The Spinner It is believed that the female spinner returns to the water at dusk to lay her eggs, and it is possibly then that the trout take notice of this pretty little fly. I have left the water very late in the evening when even the flying bats have become invisible, their presence confirmed only by their faint, high-pitched radar squeak. At such times I have heard the trout splash and feed. On what! Sedges? Or perhaps the returning Apricot spinners? When I have returned early in the morning, in the quieter bays I have seen many of these delicate spent spinners awash in the surface film.

Apricot Spinner (Price)
Hook 14
Silk Yellow
Body Polypropylene of an apricot shade
Tail Apricot hackle fibres
Hackle Apricot clipped at the bottom
Wing White raffene cut to shape and the leading edge marked with a yellow marker pen

Pond Olive Spinner (J. Goddard)
Hook 14-12
Silk Orange
Body Apricot condor herl covered with pale olive PVC
Tail Badger hackle fibres
Wing Pale blue dun hackle points tied spent
Hackle Dark honey tied in bunches under each wing

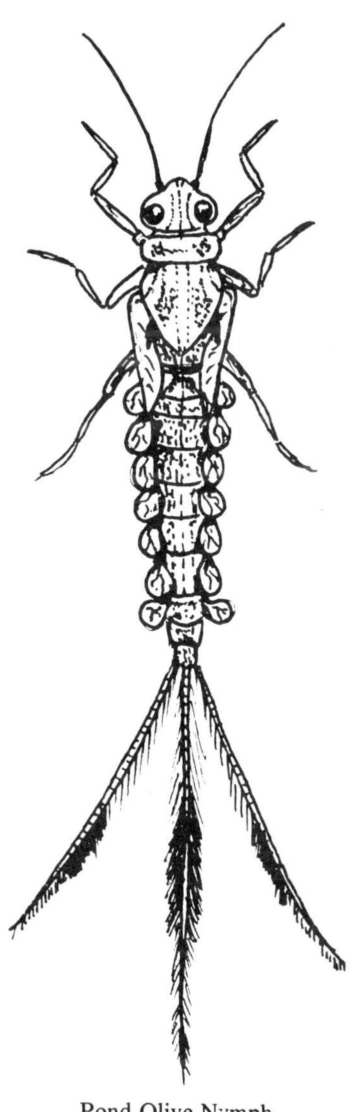

Pond Olive Nymph

The late Oliver Kite was the foremost nymph fisherman of his day. His name lives on with his most favoured of flies, the Kite's Imperial. The following pattern is his version of the Apricot spinner.

Apricot Spinner (O. Kite)
Hook 14
Silk Golden olive
Body Swan herls dyed apricot, doubled and redoubled at the thorax
Tail Pale yellow cock hackle fibres
Hackle Pale honey dun

I think I give here enough patterns for *Cloeon dipterum* for any flydresser to get his or her teeth into, and enough for the angler to ring the changes with.

Fishing the Pond Olive

The Nymph The nymph is found in the vicinity of weed beds, so it is only commonsense to fish the artificial in or as near as possible to such areas. You will catch the weeds from time to time; at least you will know you are fishing in the right place! There is also the ill wind syndrome to go with this snagging, for the disturbance of the weeds, when you pull your fly free, causes other natural nymphs and weed-clinging creatures to flee their haven, thus attracting the trout. Fish your artificial on a fairly long leader. Allow it to sink near the weeds, then twitch it back to the surface in short sharp jerks.

The Dun and the Spinner During the day try fishing the dun, in sheltered bays or coves, especially if there is surface activity. As with all dry fly fishing on still water try and anticipate the direction in which the trout is travelling, cast well ahead of the fish, and leave alone.

Fish the spinner in the same areas late in the evening. An introduction of fluorescent material somewhere in the fly dressing (e.g. as the ribbing medium) can be a decided advantage at such times.

THE LAKE OLIVE (*Cloeon simile*)

Like the Pond Olive, this insect belongs to the family *Baetidae* which includes such flies as the popular river species the Iron Blue Dun (*Baetis niger*). As a dun the Lake Olive looks very similar in size and colour to the Pond Olive already described. Kimmins, in FBA Publication No. 15 (*Ephemeroptera*), describes the wings as a mouse grey, as distinct from the Pond Olives' blackish grey. He also states that the wings are tinged with a yellow colour at the base and margins.

C. simile has a very wide distribution. It is found from eastern Kent, to Pembrokeshire in Wales, to the north-west of Scotland, and in virtually the whole of Ireland.

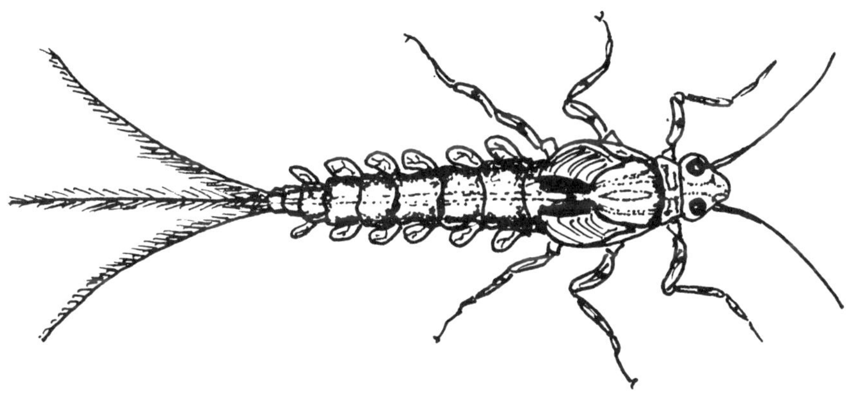

Baetis nymph

Description and Guide to Identification

The Nymph The Lake Olive nymph is about 7-10mm long. What has been said for the Pond Olive holds true for the Lake Olive, though this nymph appears to have a preference for deeper water, sheltering in weed beds near the bottom of a lake (rather than in the surface weed areas) to a depth of 3 metres.

The Dun The dun is about 7-8mm long. Two tails again, dark grey in colour. Unlike *C. dipterum* they are not ringed. Its two wings are described above. Overall the colour is a grey olive.

The Spinner The male spinner is rarely found on the water, so what follows applies to the female. At this stage in the insect's life a distinct difference can be seen between it and the Pond Olive. Whereas the female Pond Olive has an overall apricot colour, described earlier, the spinner of *C. simile* is more of a reddish brown; it also lacks the coloration on the leading edge of the wing.

Adult Emergence There is evidence to prove that this insect can have two broods a year, so it is seen on the wing from May to September.

Artificials As far as the nymph and dun stages are concerned the dressings given for the Pond Olive will serve to imitate the corresponding stages in the life cycle of the Lake Olive, so I do not propose to give dressings of my own design. Just for the record, however, I shall give some by other dressers who have thought it worthwhile to have separate dressings for the two insects.

Lake Olive Nymph (J. Henderson)
Hook 13-14
Silk Olive
Tail Blue game cock dyed light olive, tips only, the rest wound as the body
Rib Gold wire or yellow silk
Thorax Dark olive seal's fur
Hackle Two turns of blue dun hen

Lake Olive Nymph (J. T. Lane)
Hook 12
Silk Golden olive
Tail Dyed olive cock $\frac{1}{4}$in. long
Body Olive silk or nylon, tapered
Thorax Darker olive
Rib Gold wire
Hackle Olive cock hackle fibres tied under the throat and lying close to the body

On the waters that you normally fish you may have Pond Olives but no Lake Olives, or vice versa. The artificials are interchangeable. J. R. Harris, author of that fine book *The Angler's Entomology*, has a very effective pattern for the dun stage. The Lake Olive is very common in Ireland, Mr Harris's home country. Mr Harris recommends the use of the green olive pattern in spring, and the smaller brown olive version for later in the season.

Lake Olive Dun (J. R. Harris)
Hook 12-14
Tail Brown olive cock hackle fibres
Body Palest blue heron herl or white swan dyed brown olive
Rib Gold wire
Wing Dark starling tied forward
Hackle Green olive or brown olive

J. R. Harris, *The Angler's Entomology*, Collins.

The Spinner

Lake Olive Spinner (Price)
Hook 14
Silk Brown
Tail Pale blue dun hackle fibres
Body Brown polypropylene
Rib Light brown silk
Wing Either white raffene cut to shape or white hackle fibres
Hackle None

Port Spinner (C. Henry)
Hook 14-16
Silk Crimson
Tail Blue cock fibres, pale
Body Rhode Island Red hackle stalk
Wing Pale blue dun hackle points, tied spent
Hackle Light olive, two turns

Lake Olive Spinner (J. R. Harris)
Hook 12-13
Tail Rusty dun hackle fibres
Body Deep amber or mahogany seal's fur on orange tying silk
Rib Gold wire
Wing Pale grizzle dun tied spent

Fishing the Lake Olive

Generally the various stages of the Lake Olive should be fished as the Pond Olive (see p. 56).

THE CAENIS (Angler's Curse, Broadwings)

In Great Britain we have five species of Caenis; all are very similar in both form, size and colour. Identification of the individual species can only be confirmed by using a microscope, as there are certain differences in the configuration of the male genitalia. As very few anglers possess a microscope, and as the trout, keen sighted as it is, could not tell the difference, we generally treat the caenis as one fly for the purpose of fly-dressing and fishing. For the sake of completeness, however, the five species are as follows:

C. rivulorum: a river species favouring small streams
C. macrura: a common species favouring larger slower rivers

For the purpose of this book we can therefore forget those two.

C. moesta: one of the commonest of the five, found in still and running water
C. robusta: normally a stillwater species
C. horaria: normally a stillwater species

General Description and Identification

The Nymph The nymphs of the Caenis are small creatures that inhabit the silt, gravel and debris at the bottom of the water. Here they crawl slowly along the bed, often very difficult to detect because they camouflage themselves with sand and other detritus. They are usually a brown colour and in size can vary from 4 to 7mm depending on the species.

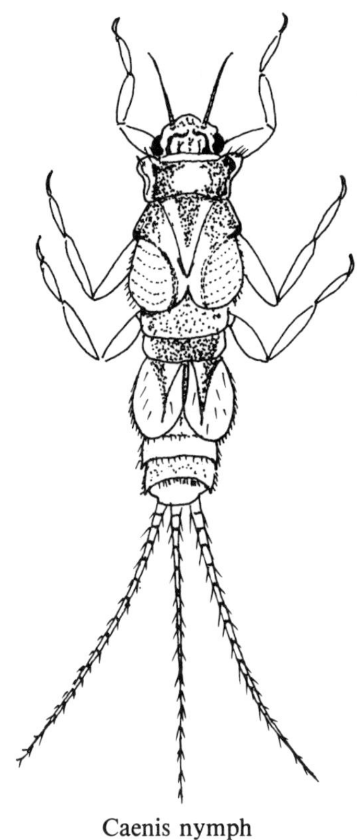

Caenis nymph

The Dun One of the most distinctive features of these insects is their extremely broad wings in relation to their body, hence one of their common names, the Broadwings. The following additional features may assist identification:

Wings: two

Tails: three, not very long

White or cream tails—possibly *C. robusta*

Grey or brown grey tails—possibly *C. moesta* (black thorax also).

As far as size is concerned, this can vary with each species, and to a certain degree within the species itself. The largest is *C. robusta*, which varies between 7 to 9mm. The other species are approximately 5-6mm.

The Spinner What has been said for the dun, or subimago, can also be said for the adult insect. The differences are mainly the much longer tails and transparency of the wings of the adult; there is little body colour change.

Adult Emergence The various species of Caenis are on the wing from the end of May until the end of August. *Caenis moesta* is a species that usually hatches out around dawn; the other two are thought to be evening flies. I have witnessed good hatches at ten a.m. and also at three in the afternoon. On both occasions I was unable to identify which particular species was causing the fish to feed avidly and causing me to curse volubly . . . it does not get named the Angler's Curse for nothing, you know!

Artificials I have never thought it a worthwhile occupation tying up flies to imitate specifically the nymphal stage of the caenis. As the nymph spends most of its life down on the bottom I have always felt it a bit of a waste of time to fish a size 18 fly along the lake bed. (There are far bigger creatures that interest the trout down there.) Fishing such a small fly in those circumstances is what I term 'fingers crossed fishing', the angling of the sublime optimist.

Small patterns of Sawyer's Pheasant Tail Nymph or Goddard's PVC Nymph can be used, not on the bottom, but to imitate the nymph as it rises to the surface prior to hatching.

Though I personally have not bothered with a nymph imitation, John Henderson gives a pattern for this stage, and fished in a manner just suggested it may well prove worthwhile. It would be a very strange sport indeed if we all stuck to our own, oft-times bigoted, sometimes dogmatic and often misconceived ideas. Part of the purpose of this book is to stimulate other anglers' lines of thought, to widen their scope, as it were; and also to give the flydresser a choice, and the angler something new to fish with.

As you can see from the coloration this artificial represents the hatching nymph rather than the nymph that crawls along the bottom. Bearing this in mind, fish it just under the surface for the best results; the trout could also take it for a swamped stillborn.

Caenis Nymph (J. Henderson)
Hook 16
Silk Flesh coloured
Tail Four or five short cock hackle fibres
Body Natural cream ostrich herl
Rib Thin gold wire
Thorax Bronze peacock herl or brown polymer dubbing

Caenis (Price)
A pattern to represent both dun and spinner
Hook 16 (or a size 14 tied short in order to increase hooking power)
Silk Black
Tail Three white deer hair fibres
Body White polypropylene dubbing
Rib Black silk
Wing White deer hair tied on by figure eight, tied spent
Hackle None

As a variant make up the wing using white Raffene that has been treated with fixative. It will have been noted that there is no hackle; the spinner rests in the surface film. This is how the trout sees it, so our artificial should be fished in the same way. For the dun use short tails. For the spinner increase the length to twice the size of the body.

C. F. Walker, in *Lake Flies and their Imitation*, gives two dressings for the Caenis spinner. He has named them the Dusky Broadwing (*C. robusta*) and Yellow Broadwing (*C. horaria*).

C.F. Walker, *Lake Flies and their Imitation*, Herbert Jenkins.

Dusky Broadwing Spinner
(C. F. Walker)
Hook 16
Tail White cock saddle hackle fibres
Body White seal's fur
Rib Silver tinsel
Wing Pale blue dun hen hackle
Hackle Short white cock

Yellow Broadwing Spinner
(C. F. Walker)
Hook 18
Otherwise, dress as for Dusky Broadwing pattern, but with a pale yellow body. C. F. Walker suggests natural silkworm silk.

Other Upwinged Flies

THE SEPIA DUN (*Leptophlebia marginata*)

I would not like to say that this fly was a truly important source of food for the stillwater trout. They are evident in many still waters, from the south of England to the north of Scotland, but never in prolific hatches like the Caenis or the Mayfly on the large Irish lakes. (The Sepia Dun is absent from Ireland.) This is a distinctive fly that cannot really be confused with any other lake species. There have been a number of artificial flies tied to represent the natural stages of this insect and I am quite often asked for them, which must indicate that on some waters they can be quite useful. On the water I fish, however, the trout appear to leave the dun and spinner well alone, and only on one occasion have I found the remains of the nymph in the stomach contents of caught fish. Another interesting fact is that they only hatch in one corner of the 15 acre lake—I have not found them anywhere else.

General Description and Identification

The Nymph Three tailed, like all members of the Ephemeroptera, the nymph does have, however, tails which are quite long and set well apart. The nymphs are an overall dark brown colour, with a darker shade of brown where the rudimentary wings are developing. The abdomen is fringed with seven pairs of gills which are plate-like, almost heart shaped, with two fine whip points from each plate. These nymphs would have been classed as laboured swimmers, but as they are quite at home slowly crawling along the bottom or clinging to weed or stones it may be sufficient to say that they are normally a slow moving species, and leave it at that. Size—about 10-12mm.

The Dun C. F. Walker referred to this fly in his book *Lake Flies and their Imitation* as 'a study in sepia', and for a broad description that cannot be bettered. Like the nymph the coloration is an overall brown; the abdomen has to my mind what could be best described as an inner orange glow which seems to shine through the prevailing sepia. The Sepia Dun has three tails, again brown, and two pairs of wings, both pairs heavily veined in brown. Size—about 10-12mm.

The Spinner There is not a great deal of difference between the spinner and the dun, male or female; the wings, as in all the species of Mayfly, tend to become more transparent, and the tails lengthen only slightly in the case of this spinner. The apex of the wings is a shade darker brown than the rest of the wing. Size—about 10-12mm.

Adult Emergence April to June, although my records show the greatest hatch in the latter half of May.

Artificials Unlike me the reader may well be lucky enough to fish a water or waters where this insect is treated with favour and savour by the trout. Then the following selection of flies could well prove their worth.

Sepia Nymph (Price)

Hook	10-12
Silk	Brown
Tail	Three pheasant tail fibres
Body	Brown dyed swan or goose fibres
Rib	Copper wire (fine)
Hackle	Palmered brown hackle, shortish fibre
Thorax	Brown seal's fur
Wing case	Crow or jackdaw wing slip

Sepia Nymph (R. Walker)

Hook	12-14
Silk	Black
Tail	Sepia dyed swan or goose
Body	Brown ostrich herl
Rib	Black floss
Thorax	Black floss
Wing case and legs	Sepia brown swan or goose

One of the patterns for the nymph I am constantly asked for is the Sepia Nymph devised by Richard Walker—this fly is a good fish catcher.

Of the many books on fishing, flydressing and angling entomology that have come from the USA over the last few years, one that stands out as a unique work is the book *Nymphs*, by Ernest Schwiebert. His book is monumental in its concept, dealing with most of the aquatic stages of the American angler's natural flies, and there are plenty, believe you me. Amongst the nymphs of the American Ephemeroptera there are some closely related to our own *Leptophlebia*; the following dressing is based on a pattern of Schwiebert's tied to represent *Leptophlebia cupida*, one of the smaller American members of this family. Ernest Schwiebert gives the common name for this nymph as Early Black Quill.

Leptophlebia Nymph
Hook 12-14
Silk Brown
Tail Dark wood duck fibres (use brown mallard)
Body Blackish brown hare's mask
Gills Brown marabou either side of the body tied down with fine gold wire (Schwiebert uses olive marabou for his Dark Quill pattern)
Thorax Blackish brown hare's ear
Wing case Medium brown feather
Hackle Medium olive hackle fibres
In Schwiebert's original dressing he also gives a head of brown nylon.

So much for the nymphs. The next stage has been fairly well imitated by flydressers, and the dun is a particularly satisfying fly to tie.

Sepia Dun (Price)
Hook Up eye 12 or equivalent Yorkshire flybody hook
Tail Three brown mallard fibres
Body Dark brown flybody fur (or seal's fur)
Rib Orange brown silk or nylon
Wing Dark partridge hackles set upright
Hackle Brown cock

Sepia Dun (O. Kite)
Hook 14
Silk Dark brown
Tail Dark brown or black cock hackles
Body Dark brown dyed heron herls, doubled and redoubled at the thorax
Rib Gold wire
Hackle Blackish brown cock (or brownish black, whichever way you look at it)

Sepia Dun (D. Jacques)
Hook 12
Silk Maroon or claret
Tail Three pheasant tail fibres set well apart
Body Pheasant tail fibres
Rib Fine gold
Wing Mottled brown cock wing slips
Hackle Furnace

The late Oliver Kite evolved a pattern for the Dun stage—it is as given above. As an alternative I would like to give one of David Jacques' flies. His precise attention to entomological detail is reflected in the flies he uses. David is one of the few purist anglers I have had the privilege to meet.

The Spinner

Sepia Spinner (Price)
Hook 12
Silk Brown
Tail Brown mallard fibres
Body Brown flybody fur
Rib Claret silk
Wing Bunches of partridge spider, tied spent
Hackle None

Swisher and Richards, the American angling entomologists, give in their book *Selective Trout* a number of direct representational patterns for many of the American Ephemeroptera. (They are well known for their series of No-hackle and Parachute flies.) The following pattern is based on their version of *Leptophlebia cupida*, the Black Quill, which in some areas of the USA is known as the Whirling Dun.

Leptophlebia Spinner

Hook	12-14
Tail	Dark blue bronze dun, set well apart
Body	Dark reddish brown seal's fur
Wing	Blue/bronze hen tips tied spent
Hackle	None

My final choice of spinner is yet again one of C. F. Walker's, and it is his version of this lake dwelling fly.

Sepia Spinner

Hook	12-13
Silk	Brown
Tail	Dark grey or black cock saddle hackle fibres
Body	Dark brown seal's fur with yellow seal's fur worked in
Rib	Gold tinsel
Wing	A ginger and grey cock hackle, tied spent, sloping aft
Hackle	Dark brown honey dun or none

Fishing the Sepia Dun

Try the nymph slowly along the bottom in the month of April, especially if you know that the natural insect occurs on your water. When the duns emerge fish the nymph near the surface to imitate the natural as it is about to hatch. As far as the winged stages are concerned, use them if the fish appear to be taking them, and fish them in the same way as any other dry fly described earlier.

THE CLARET DUN (*Leptophlebia vespertina*)

As you can see this insect is a close relation of the Sepia Dun. The word vespertina comes from vespers—evening—though the fly hatches at other times also. (Incidentally, in *L. marginata* 'marginata' means 'lined'.)

This fly is fairly well distributed and unlike its cousin it is found in fairly large numbers in Ireland. It appears on more acid waters than those frequented by the Sepia Dun, and is often found on high mountain lakes. It also has a longer emergence period than the Sepia Dun.

General Description and Identification

The Nymph At a glance there is very little to choose between the nymph of the Claret and the Sepia; there is, under a strong magnifying glass, a noticeable difference in the shape of the breathing gills.

The Dun The dun has three tails and two pairs of wings. The wings of the Claret Dun are far less deeply marked than the Sepia's; the fore wings are a dark grey and the hind wings a buffish grey. This is one of the main identification points, and is visible to the naked eye even from a distance. The body, though a darkish brown, has a tinge of claret about it—hence the insect's common name.

Sepia Nymph gill   Claret Nymph gill

The Spinner No great difference, except for the more transparent wings. The Claret spinner's wings lack that darker colour on the leading edge that the Sepia spinner possesses. The tails are more distinctively ringed. Size—about 10mm long.

Adult Emergence May to end of June.

Artificials

Nymph Use the patterns given for the Sepia Dun.

Dun

Claret Dun (Price)

Hook	14
Silk	Claret
Tail	Bronze mallard fibres
Body	Tail end brown/claret polypropylene, darker brown for rest of fly
Rib	Maroon silk
Wing	Blue dun hackle points, set upright
Hackle	Black and claret mixed (two hackles)

Claret Dun (J. R. Harris)

Hook	14
Silk	Claret
Tail	Dark blue dun cock
Body	Dark heron dyed claret, or mole and claret mohair spun on claret silk
Rib	Fine gold wire
Hackle	Dark blue dun clipped underneath to form V

Claret Dun (J. Henderson)

Hook	13
Silk	Dark claret
Tail	Rusty dun cock spade hackle fibres
Body	Dark claret seal's fur
Rib	Fine gold wire
Hackle	Rusty dun cock

Spinners

Claret Spinner (Price)

Hook	14
Silk	Claret
Tail	Bronze mallard or dark black/claret hackle fibres
Body	Dark claret and brown polypropylene dubbing
Rib	Yellow silk
Wing	Treated raffene grey colour, tied spent
Hackle	None

Claret Spinner (J. R. Harris)

Hook	14
Silk	Claret
Tail	Dark blue dun cock
Body	Dark claret seal's fur
Hackle	Blue dun or rusty dun cock

The next artificial spinner is, strictly speaking, a variant of the Sherry Spinner, the common name for the spinner of the Blue Winged Olive. The dressing is to be found in the *Further Guide to Fly Dressing* by John Veniard, and is a fly tied by the Eastbourne flydresser Peter Deane, who to my mind ties some of the best dry flies fished on the southern chalkstreams. The fly is named the Claret Sherry Spinner (a mixed wine fly if ever there was one), and the dressing will serve to imitate our Claret Spinner too.

Claret Sherry Spinner
(P. Deane)

Hook	14
Silk	Claret (not given in *Further Guide to Fly Dressing*)
Tail	Three grizzle hackle fibres (for this fly the darker the better)
Body	Lightly spun claret fur or wool
Hackle	Grizzle

To close this piece on the Claret Spinners, mention had better be made of the spinner of the Iron Blue Dun, the small dark fly of the rivers. This stage in that insect's life is also often referred to as the Claret Spinner. I hope that this will not serve to confuse the reader.

Fishing the Claret Dun

Fish all stages in exactly the same way as the Sepia Dun (see p. 66).

THE SUMMER DUN (*Siphlonurus* species)

The main reason for my including this fly is that on two day ticket waters in Sussex, namely Darwell and Powdermill, the fly hatches out in quite large numbers. In Britain we have three species of Summer Dun, all of them very similar in appearance. They are:

S. linnaeanus: most common in Ireland

S. lacustris: found on small lakes and tarns, often at altitude, also found in the slower reaches of small streams. (I have found this species on the Cherry Brook in the middle of Dartmoor.)

S. armatus: found on lakes and streams of a 'chalky' nature.

The species which I collected on the two reservoirs mentioned, I identified as *S. armatus*, and John Goddard kindly confirmed my identification. As it happens it is the rarest of the three species. The word *armatus* ('armed') is attributable to the fact that the insect has a sharp spine on the ninth segment of its abdomen.

General Identification and Description

For the purposes of fishing and flydressing all three species can be considered as one and the same, for they are all very similar in appearance. The fly is best described as a medium to large Mayfly.

The Nymph A large nymph, approximately 19mm long, with gills that are quite large running the length of the abdomen. The colour of the nymph is a dark olive and it is a fairly active creature.

The Dun The dun has four wings and two tails, and is about 14mm in length (excluding tails). The abdomen is a brownish olive with distinct triangular serrations of yellow olive running along the sides. The wings have a distinct greenish olive hue.

The Spinner Very similar in appearance to the dun but with a marked increase in the length of the tails and a greater transparency (without the olive tinge) to the wing.

Adult Emergence From May to August. On the reservoirs mentioned I have witnessed good hatches during the latter half of May.

Dressings

Summer Dun Nymph (Price)
Hook L/S 12-14
Silk Brown
Tail Brown olive fibres
Body Brownish olive seal's fur
Rib Gold wire and palmered olive hackle
Thorax Olive and yellow seal's fur mixed
Wing case Brown feather fibre
Hackle One turn of dark olive hen or cock hackle

The Summer Dun (Price)
Hook 12
Tail Dark olive cock hackle fibres
Body Brownish olive seal's fur or polypropylene dubbing
Rib Yellow silk
Wing Partridge dyed pale olive
Hackle Brown olive and light olive mixed

Summer Dun Nymph
(C. F. Walker)
Hook L/S 12
Tail Brown partridge
Body Brown ostrich
Rib Silver
Thorax Dark hare's ear or brown seal's fur
Hackle Medium grizzle honey

Summer Spinner (Price)
Hook 12
Tail Long dark olive fibres
Body Polypropylene dyed brown (light)
Rib Yellow silk or nylon
Wing Light partridge, tied spent
Hackle None, or a dark olive clipped at the bottom

Fishing the Summer Dun

If you are lucky enough to fish waters where this fly is in evidence then any of the patterns given above should suffice. I found that the insect larvae like to be very close to the bank; in water up to three feet I have collected specimens within six inches of the shore line, and I have not found any further out in deep water. So fish your nymphs close to the bank, and fish the dun and spinner in the usual way, but again close to the margins, for this is where the fly hatches.

THE BLUE WINGED OLIVE (*Ephemerella ignita*)

Quite often Ephemeridae whose usual habitats are rivers find themselves on still water. They are either windborne, or they have been carried there on the streams that feed the lakes. These visitors from the moving water, if they occur in significant numbers, will often be taken by the trout. One of the most common of these flies is the Blue Winged Olive, and I have often seen these on the water I fish. As a point of interest the Blue Winged Olive is the only British olive to have three tails, a good aid to identification. It is also one of the most prolific and common species.

As this is not strictly speaking a stillwater fly I shall only give one dressing for the record's sake. In my opinion one of the finest dressings for the Blue Winged Olive is the pattern devised by David Jacques; it is as effective as it is realistic.

For the spinner, what better pattern than the dry Pheasant Tail or the Lunn's Particular, two traditional flies that continue to catch fish in these days of modern materials and new ideas? A good fly is a fly for all time, not just a passing fancy. As an alternative, use any named Sherry Spinner pattern.

Blue Winged Olive (D. Jacques)
Hook 14
Silk Orange
Tail Light olive hackle fibres
Body Dirty yellow ostrich herl with olive dyed PVC over
Wing Coot (these are tied upright but with a slight backward inclination)
Hackle Light olive

Lunn's Particular
Hook 14
Silk Brown
Tail Natural red cock hackle fibres
Body Natural red hackle stalk
Wing Medium blue hackle points
Hackle Medium Rhode Island Red

Pheasant Tail Spinner
(G. E. M. Skues)
Hook 14
Silk Orange
Tail Honey dun cock hackle fibres
Body Dark pheasant tail fibres
Wing None
Hackle Rusty dun

THE YELLOW MAY (*Heptagenia sulphurea*)

At certain times the olive surface of a lake is brightened by a medium sized Mayfly that is bright yellow in overall coloration. The fly in question is the Yellow May. As far as our lake is concerned, I believe the few I have observed have come in on the chalkstream feeder, but in other areas very large numbers are known to hatch out. Some of the Irish limestone lakes are such places, and there the fly is sometimes called the Yellow Hawk. Generally speaking, if only a few of these sulphur coloured flies hatch out the trout seem seldom to take them, but if the numbers on the water resemble yellow confetti then the trout will feed as avidly on them as they will on any insect. I have had this experience in Devon on the river Torridge, in June, when yellow flies were the order of the day. The Yellow May has a long emergence period, lasting from May right through to September. As for a pattern, you might like to tie the following. It looks like a small canary, but it does work during those mellow, yellow times.

Yellow May (Price)
Hook 12-14
Silk Yellow
Tail Yellow hackle fibres
Body Yellow dyed swan or goose
Rib Pale green terylene
Wing Two swan or goose slips, or partridge dyed yellow
Hackle Yellow

This sunset yellow of a fly concludes the chapter on the beautiful Mayflies (or Dayflies, as they are sometimes called). I have already said that they may not be of such importance to the stillwater angler as other less attractive creatures, but on occasions they are, and they are always worth looking and marvelling at, anyway.

Crustacea

THE FRESHWATER SHRIMP (*Gammarus pulex*)

The preceding parts of this book have dealt with insects of one sort or another. We have now moved on to a different class of creature, namely Crustacea, and in particular the Freshwater Shrimp (*Gammarus pulex*). This so-called shrimp is not a shrimp at all, but belongs to an animal group called *Amphipoda*. Perhaps the nearest relative of our shrimp is the drab sandhopper that fandangos on the beach whenever we lift up damp seaweed. There are a number of other Amphipods often referred to as 'shrimps', but these species are to be found either in wells and springs or in high altitude lakes. These other species are thought to be Ice Age relics.

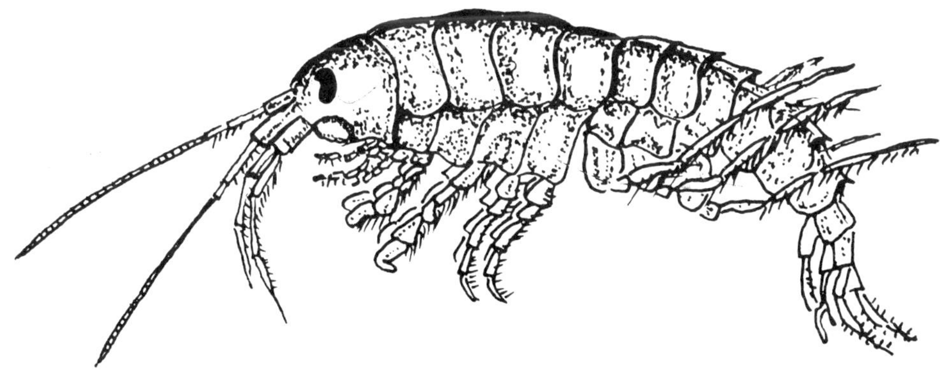

Freshwater shrimp (*G.pulex*)

Gammarus pulex is the angler's shrimp. It is found in all waters, from chalkstreams to man-made reservoirs, it can grow to the length of 20mm and, unlike many of the insects that interest the fisherman, there is no defined mating period. The shrimp mates throughout the year, as those who have the creatures in an aquarium can testify.

What of colour? Well, there is a wide range of colours in the natural, ranging from drab dark olive to an almost bright yellow. Some specimens I have taken have been a greyish white. Some believe the shrimp turns an orange shade during mating. I have not found this to be a general rule, and shrimps of all shades and colours seem to indulge in the usual routine.

General Description and Life Cycle

I do not suppose that there can be many anglers who have not seen the freshwater shrimp. To repeat, it belongs to the Amphipoda ('amphi'—both ways; 'podos'—foot), so called because they have two different types of legs. In appearance the *G. pulex* is a multi-legged, flattened creature with the whole of the body a segmented carapace. At rest the 'shrimp' is hump-backed, but when swimming it extends its body and in a rapid jerking motion it propels itself through the water.

The shrimp is the scavenger of the water, eating any dead animal or vegetable matter. They have been known to prey on live nymphs and on occasions they have been observed to be cannibals—charming little creatures. *G. pulex* swims on its side. Some flydressers tie their flies with a weighted back to allow their artificials to swim upside down, under the misapprehension that the natural swims on its back. It does not matter where you weight the fly, and to be perfectly accurate we should in fact weight one side only, but what nonsense! I have caught too many trout on shrimp patterns weighted with lead wire wound around the shank not to realise the trout don't bother to look which way up the fly is.

The female shrimp can lay anything up to 1000 eggs, depending on the age of the crustacean and the prevailing habitat and food supply. The eggs are carried by the female in a brood pouch, and the infant shrimp looks exactly like the adult, for there is no larval form. Like most invertebrates *Gammarus* undergoes a series of skin moults in order to grow. I believe *G. pulex* has ten skin moults. The male shrimp is larger than the female, and it can be seen clasping the smaller female shrimp beneath its body and will swim with her in this position for days. In one book I have read that the female is the larger and the male carries her on his back; this is not true. Again, another fallacy found in some angling books is that there is a definite mating period. There is not; they mate all the year round . . . cosy. In his book *Pond Life* (Burke), W. Engelhardt states that under favourable conditions the *G. pulex* can develop in large numbers, anything up to 400 in a square metre. It goes without saying that any creature of the size of our freshwater shrimp (and in the numbers in which they occur) must form an important source of food for the brown and rainbow trout we seek.

W. Engelhardt, *Pond Life*, Burke.

Artificials There are a number of flies tied to represent this creature. The following patterns, mine and other flydressers', are but a few that can be found within the pages of modern books on flydressings. As far as catching fish is concerned I should like to think that my patterns were surefire killers, but alas it is not true. Mine are no better and no worse than any of the other dressings I shall give.

The first is a pattern that I first had published in 1968 in the journal *Angling*. It was one of a series utilising deer hair as the main material in the fly's construction.

The next two shrimps I have based on an American original pattern, and it differs from many other patterns in that the lead wire used in the fly is wound as a rib after the dressing, so that it serves two purposes. If, after tying the fly, you gently squeeze the back of it, the fly flattens to a natural shrimp shape. The first of the two patterns is the Golden Shrimp, and it is supposed to imitate a natural shrimp just after it has shed its old skin. I found (by tank observation) that such shrimps are very much lighter than others swimming around, and were the first to be taken by the fish, for against the dark green weed they stuck out like sore thumbs.

Deer Hair Shrimp
Hook 14-10
Silk Black or orange
Body Underbody of lead or copper wire, then grey deer hair spun on. This deer hair is clipped smooth top and bottom leaving unclipped hair at the bottom to imitate the shrimp's legs.

Golden Shrimp (Price)
Hook 14-10
Tail A downward pointing tuft of golden olive hackle fibres
Body Golden olive seal's fur
Hackle Palmered golden olive cock clipped top and sides
Back Either polythene, PVC or yellow dental dam
Rib Lead wire (a yellow terylene can be used instead, but weight the hook first in the conventional manner)

Olive Shrimp (Price)
Dress exactly the same as the Golden Shrimp, but use a dark olive seal's fur and hackle.

The Shrimper is a pattern devised by John Goddard, and of all the artificial shrimp flies I am asked to tie, this is the most popular. The following pattern is the Otter Shrimp, devised originally by Ted Trueblood. The next pattern is the creation of one of the most creative flydressers in the United States, Dave Whitlock, whose flydressing and book illustrations are world famous. For this pattern the hook must be bent in the shape shown in the illustration.

The Shrimper (J. Goddard)
Hook Down eye 10-14 Limerick (weighted)
Silk Orange
Body Fine copper wire wound in thickly at the centre to form a hump, covered with olive brown seal's fur
Hackle Olive cock
Back PVC, cut wide in the centre and tapering at each end

Another version of the above has a little orange fluorescent silk wound over the body material.

Freshwater Shrimp (C. F. Walker)
Hook 10-14
Body Olive, brown and amber seal's fur
Rib Gold tinsel
Hackle Pale watery olive hen hackle

Otter Shrimp (Ted Trueblood)
Hook 10-14
Silk Olive
Body Grey otter fur (substitute water rat or mole) and a small amount of brown seal's fur
Hackle & Tail Grouse body feather (the fly can be weighted)

Dave's Shrimp (D. Whitlock)
Hook 8-14, bent to shape
Silk Olive
Tail Barred lemon woodduck feathers
Body Synthetic fur yellow and olive, bleached beaver belly (natural seal substitute) and muskrat belly fur
Hackle Tied at the throat, barred lemon woodduck

Dave's Shrimp

Fishing the Shrimp

First of all, when to fish this imitation? Well, the natural *Gammarus* is active right through the fishing season, so there is no reason why it should not be fished from the early months until the close. When there is no obvious surface activity and the angler is at a loss what to try, then the shrimp could be the face saver, for if the trout are not tempted to rise to surface or surface film food then they could be feeding somewhere near the bottom on something. Try a shrimp.

Fish the shrimp near to weed beds and known sunken features such as submerged trees, hedgerows, and roadways. The fly can be fished on both sunk and floating lines, or with a sink tip. This last method can be very effective over weed beds, the floating line resting on the weed and the tip sinking into the water over on the other side. In such circumstances fish very slowly, watching the floating portion, where it enters the water, for the slightest movement. The shrimp is not the slow push-over that many will have us believe, for when disturbed it can move at a fair rate of knots. So, if fishing with sinking or floating lines, vary the speed of retrieve.

THE WATER LOUSE (*Asellus aquaticus*)

The Water Louse, Hog Louse, Water Slater and Sow Bug are all names for the little creature *Asellus aquaticus*, a member of the family *Isopoda*, and, like the Freshwater Shrimp, a crustacean.

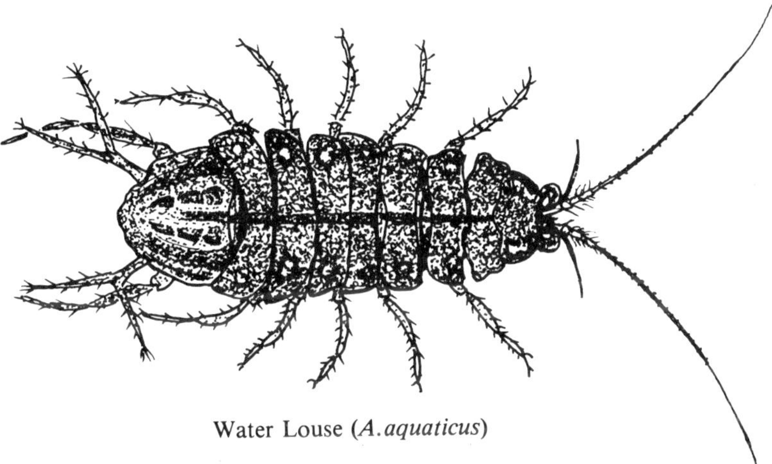

Water Louse (*A.aquaticus*)

Isopoda ('iso'—the same; 'podos'—foot) is a name which is attributable to the fact that, unlike *Gammarus pulex*, all the creature's legs are the same. There have been three recorded species of water louse in Great Britain, the most common being *A. aquaticus* and *A. meridianus*. The third is a cave dwelling species found in the south and, like many cave dwelling animals, it has no eyes. Unless there are some troglodyte anglers, they are of little consequence to the fisherman or fly-dresser. (I must admit that, on reflection, I have seen reservoir anglers who have looked and behaved very much like cavemen.)

The Water Slater (or Louse) and the Shrimp share the same habitat, though I have found that in my many insect forays to river, pond and lake, the louse seems to favour more muddy areas, and has a greater survival factor in polluted or stagnant water. They grow to a size of 12mm, and the male is the larger. The overall coloration is a dirty olive grey.

General Description and Life Cycle

The Louse has seven pairs of legs, and has a flattened, well segmented body. The antennae are long, almost the length of the body, and the tails are branched, with a distinct fork to each tail. For the *G. pulex* I drew the comparison between it and its near relative the sandhopper. Well, a close relative of the Water Louse is the Wood Louse, but unlike the Wood Louse or Pill Bug the aquatic cousin does not roll up into a ball.

Another thing these creatures have in common with our freshwater shrimp is the fact that they are scavengers, feeding mainly on decomposing organic matter and algae. Though they can swim, they have not the speed of the shrimp; they seem to prefer either to tip toe along the bottom, or clamber in and out of the weeds.

During mating the female is carried around beneath the male. She carries her eggs in a mass, beneath her body, by means of plates specially developed for this purpose at the bases of the first four pairs of legs. The eggs vary in colour from a greyish white to a pale green. Like the freshwater shrimp, this pairing together during mating lasts for about a week, and like the shrimp mating can take place at any time during the year.

When the eggs hatch out the infant lice hang onto the female's legs for some time before making their own way in the world.

Artificials

When observing creatures in my aquarium I have noted that the unfortunate louse was always eaten by fish or other predators before the shrimps. This was, I suppose, because the louse was just not quick enough in evading its enemies, it did not swim fast enough. This observation indicates to me that any artificial should be fished slowly.

Water Louse (Price)

Hook 12-14
Silk Grey or brown
Body Grey wool over lead wire and flattened
Tail Grey partridge hackle fibres
Hackle Blue dun or grey wound palmer, clipped top and bottom
Back Grey mallard tied at the tail, taken over the back and tied at head, then varnished

The next pattern represents a female louse carrying eggs, a first course meal for the trout complete with an hors d'oeuvre of caviare as it were. It is the creation of Ann Douglas of Footscray, Kent, one of the few creative women flydressers we have in the country.

Freshwater Louse (A. Douglas)
Hook 12-10
Silk Brown
Back Brown mallard flank
Body In three parts: the first part is hare's ear and mole mixed; the second part is green raffene to represent the egg sac; the third part, hare's ear and mole mixed.
Legs Brown partridge hackle
Feelers Brown partridge

Dressing Instructions
After winding the tying silk to bend of hook, tie in a strip of brown/grey mallard flank, then a partridge hackle by the tip. Dub onto the silk the hare and mole fur and apply to the hook. Tie in the strip of raffene and dampen, twist this damp raffene around until it forms little knots and, still keeping a tight hold, wind around the hook shank. This is the egg sac. Now dub on the remaining third of hare and mole. Tie in a few partridge fibres over the eye of the hook, then take the partridge hackle over the back and tie off at the head. Follow this with the mallard feather and tie off, whip finish and varnish; for durability varnish the back and also the raffene egg sac, and the fly is complete.

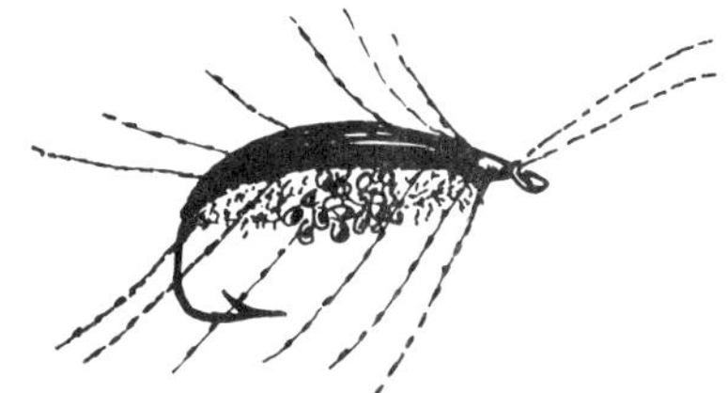
Douglas Water Louse

C. F. Walker, in *Lake Flies and their Imitations*, gives one of the few British patterns for this creature. As far as British flydressing is concerned the Water Louse has been virtually ignored. There are many more American patterns for the Water Slater than we have for our indigenous species. One such fly is the pattern tied to represent *Asellus communis*; one of the more common American lice, it is the creation of Ernest Schwiebert, whose book *Nymphs* was mentioned earlier.

Water Louse (C. F. Walker)
Hook 12-14, weighted
Body Grey brown hare's ear flattened horizontally
Rib Silver tinsel
Hackle Grey partridge wound half way up the body

American Sowbug (E. Schwiebert)
Hook 14-16
Silk Grey nylon
Body Grey muskrat
Thorax Dark grey muskrat
Legs & Swimmers Dark pheasant fibres
Antennae Dark grey fibres

Fishing the Water Louse

Just like the shrimp, the water louse is in evidence throughout the year. It is most active around weed beds, and anywhere where there is sunken vegetation. Any log or tree root that is pulled out of the water usually has a great number of lice attached. Fish your artificials close to weed beds, either with a floating line and long leader or with a sunk line. Fish slowly, inching your fly back, for the best results.

THE WATER FLEAS (*Daphnia species*)

Among the many and varied creatures that swim in our lakes and reservoirs are a number of small crustacea that figure highly in the diet of the trout. Some of these are so small as to pass notice, and may well be taken by the trout accidentally. Creatures such as the minute Ostracods are such crustacea; they resemble tiny beans with retractable legs, and they are far too small to imitate. Another very important family of crustacea which can, if we magnify our imagination, be imitated by the flydresser are the Water Fleas, members of the family Daphnidae.

The daphnia are perhaps one of the most important links in the ecological food chain of a reservoir. They are preyed on by a wide variety of aquatic animals; where the daphnia are, there you will find the trout.

Daphnia

There are many different species of so-called water fleas. Some are only found in the margins or in weed beds, others are free swimming in the lake or reservoir, forming an important part of the water's zooplankton.

General Description and Life History

Some of the more common members of this group are as follows:

D. pulex	approximate size 1mm
D. magna	5mm
D. longispina	2mm
D. cucullata	1.5-2mm
D. hyalina	2mm

All the above are found in a wide variety of waters, from small ponds to large, man-made reservoirs such as Grafham Water.

The daphnia vary in colour from a watery olive green to a pinkish orange. This latter colour is formed in exactly the same way as the red colour in the bloodworm, from haemoglobin. The haemoglobin is more evident in daphnia from stagnant waters, or at least in those from waters with a low oxygen content. It is there to help the daphnia make the most of what oxygen is available.

Most daphnia are constructed in the same way, with two pairs of antennae, which they continually wave about. The second pair of antennae are their means of propulsion. Most of the daphnia are transparent, and the observer can see all the vital organs if he uses either a microscope or high powered lens. (The colour plate depicting the daphnia was taken using both extension tubes and a fully extended bellows.)

The legs of the daphnia are fringed with hairs, and are concealed within the shell of the creature. These legs are used as food filters, filtering the algae which forms the main part of the diet of these minute crustacea. Another interesting attribute of the daphnia is their ability to change shape with the seasons. Many of the species have two distinct forms, a summer form and a winter form. During the winter months they appear to grow what can be best described as a horn from the top of the head. The daphnia see by means of a compound eye, which is very noticeable when viewed under a glass.

Nature always seems to compensate in the case of preyed on animals. The prey usually occur in large numbers, as, for cxample, some of the antelope and gazelle in Africa, or the large shoals of fish that are pursued by the more predacious species. So too, in the insect world, creatures that form an important part of the diet of other creatures usually exist in large numbers to compensate for that fact. In some cases, such as the aphids, nature further aids them by the process of parthenogenesis, a long word simply meaning virgin birth. Furthermore, whole broods of live, entirely female aphids are laid. This virgin birth and viviparous habit is emulated by the daphnia. The eggs are kept in the space between the abdomen and the carapace, are easily seen, and in the summer months quickly develop into perfect miniature copies of the adult. These females are ejected by the parent and after about three moults they are considered to be sexually mature and the process continues. After a few generations male daphnia are born. As the summer draws to an end the female starts to lay other eggs, which require male fertilisation. They are known as the winter eggs, and remain dormant throughout the winter before hatching into female daphnia. So life goes on. It does not require a great deal of imagination to appreciate that, were it not for such predators as fish and insect larvae, the world would long ago have drowned under a sea of daphnia.

Artificials

Two or three years ago I wrote in *Trout and Salmon* of an experience of Daphnia-feeding fish on a lake I was fishing. I had tried everything in the book, to no avail, for I did not know at that time what the fish were taking. Out of sheer desperation I put on a small Partridge and Orange, a fly more at home on some northern beck and not on a southern lake. Bang! first cast with the fly, a good rainbow trout. Out of the mouth spewed forth a veritable jam of daphnia. I told another friend what they were eating, and soon he was into a fish. His fly was a Grenadier nymph, again a fly of an orange colour.

The following fly I devised for such daphnia-feeding occasions, and I tie it down to size 16. Any smaller and you would lose the fish, but that said I have found size to be relatively unimportant. A size 14 is taken avidly. I believe the trout go through the shoals of daphnia with mouth open, and your fly is just taken as either a larger-than-normal daphnia, or a small group of daphnia. This may account for the success of such lures as the Whisky Fly which, when used in a shoal of daphnia to daphnia-feeding trout, has proved successful time and time again.

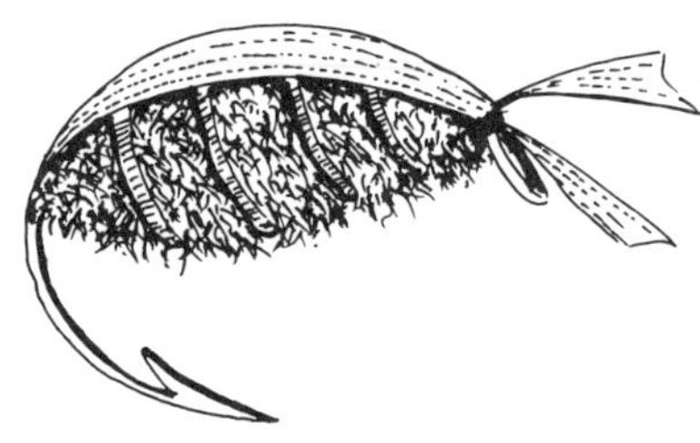

The Orange Nymph (Daphnia)

Hook	16-14
Silk	Orange
Body	Pinkish orange seal's fur
Rib	Gold tinsel
Back & Antennae	Orange dyed swan or goose

Fishing Daphnia

Daphnia are sensitive to light; the more intense the light the deeper the shoals will go. The zooplanktonic layer moves back up to the surface as the light fades. This information should indicate where and when to fish the artificial. I now fish the Orange Nymph as a dropper with a much larger nymphal pattern (or lure). For this I usually use a floating line and long leader, and retrieve the flies at varying speeds at varying depths until a take (I hope) occurs. On some occasions the fish can be seen actively feeding on the browny orange masses of daphnia as they hop, skip and jump under the water. Then one would cast in the vicinity of the fish.

Corixa

ORDER: Hemiptera

Back with the insects once more, we come to an important little creature for the lake or reservoir angler, the Lesser Water Boatman. There are 32-33 different species of this insect family in Great Britain, and they are found in all types of water, even brackish estuarine waters. Imitations of this active little insect can prove very killing on most stillwaters, but it is on the shallower, put and take fisheries where they come into their own.

They belong to a group of insects known as the Bugs or True Bugs, a number of which are aquatic. Amongst the aquatic bugs we have the Water Measurers, Pond Skaters, Water Scorpion, and Water Stick Insect. As for the terrestrials, they include such creatures as the Squash Bugs, Leaf Bugs, and Assassin Bugs.

General Description and Life Cycle

The corixidae spend both larval and adult stages in water, though on some occasions the insect will fly to a neighbouring water, just for a change I suppose, or maybe because of an overpopulation problem. I have often taken specimens from my moth trap in the garden.

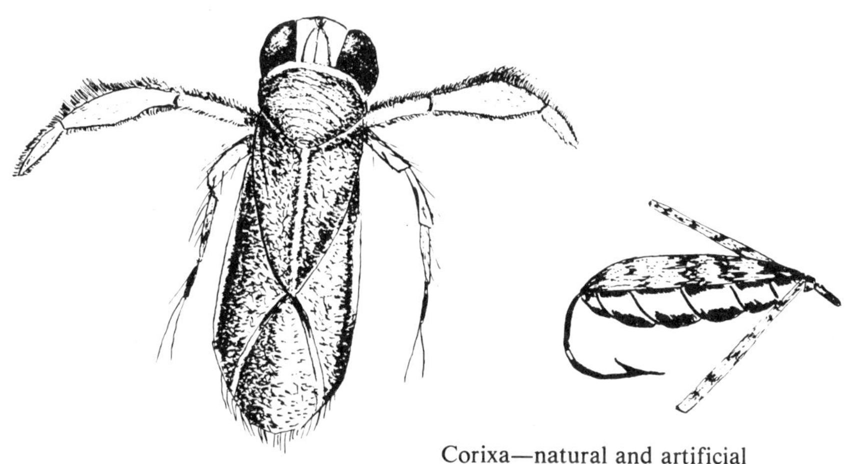

Corixa—natural and artificial

They are air breathing; they can be seen darting to the surface for air, which they take in through spiracles on their prothorax. This air is held under the wingcases and on unwettable hairs on their abdomen. Because of this built-in air bubble they are much lighter than water, and therefore they have to cling on for dear life to the bottom, or onto any weed or stone they rest on. Their main diet consists of algae and other plant detritus, and generally they are not predators. They are very fast and efficient swimmers, and unlike their big brothers the Water Boatmen they swim the right way up. Another interesting feature is their ability to striate, that is to say they make an audible chirruping in much the same way as a grasshopper or cricket. One species is known for this chirruping, and has been named the Water Singer for this reason. It would appear that only the males have this ability. The corixa vary in size from about 3mm to quite large species of 13mm. One of the most common of the corixa is *C. punctata*; this particular water bug is found all over the British Isles including Ireland. It is one of the few day flying species, and it will move home if the conditions in its original habitat become untenable for any particular reason.

Mating takes place usually in January and February, and eggs are laid at this time on convenient submerged vegetation. Depending on the prevailing temperature the eggs will hatch out. Southwood and Leston, in their book *Land and Water Bugs of the British Isles* (Warne), state that the temperature must be a constant 14°C for the eggs to develop in about 20 days. The nymphal stages of the water bugs are known as instars, and with each moult a different form of the nymph develops. In the case of *C. punctata* there are five such instars, with a total developing time of about 2-3 months until adulthood.

T. R. E. Southwood & D. Leston, *Land and Water Bugs of the British Isles*. Warne.

Artificials

There are a number of effective 'flies' dressed with this insect in mind. All of them catch fish. Whilst observing the corixa in my aquarium, I was impressed with the speed at which it darted to the surface from the bottom after a period of comparative inactivity. I was further impressed by the silver appearance of the insect as it flashed back to the bottom. With its built-in air supply it looked like a fast drop of mercury. This silver appearance prompted me to tie the following pattern. It subsequently

Silver Corixa (Price)
Hook 16-10
Silk Black
Body Flat silver lurex over a light silk
Rib Oval silver tinsel
Back and swimming legs Cock pheasant tail fibres
The fibres are taken over the back and divided at the head, most are then cut off, leaving two for paddles. The back is then varnished for durability.

emerged that Col Joscelyn Lane had made the self-same tank observations, and had evolved a pattern almost identical to mine; the trouble is, he created his before I did.

Alternative Patterns

Having just mentioned Joscelyn Lane's pattern, I had better give his dressing. It does not vary very much from mine, but if anything mine is somewhat easier to tie. Richard Walker has also devised fish catching corixa patterns. One of his is the Green Corixa. Many of the infant instars of the Corixids are in fact a green colour, as is the final instar of the Water Boatman proper (*Notonecta glauca*). This pattern of Mr Walker's imitates these larval forms well. Richard Walker's brown version imitates the larger of the corixa species. The dressing is as follows.

Streaked Corixa (J. T. Lane)
Hook 12
Silk Golden olive
Body Strips of raffia over which is wound flat silver tinsel
Back Brown partridge
Paddles A small red cock hackle with the fibres clipped down

Green Corixa (R. Walker)
Hook 14
Body White floss (fat)
Rib Silver
Back & legs Olive dyed swan, dyed a pale shade

Large Brown Corixa (R. Walker)
Hook 10
Body White floss
Rib Silver thread
Back & legs Brown speckled turkey

In *Fly Dressing Innovations* (Ernest Benn) by the ubiquitous Richard Walker there is to be found a most interesting version of a corixa fly credited to David Collyer. It is the Plastazote Corixa. Plastazote is a plastic, an expanded polyethylene. Incorporated into flydressings it gives an extra buoyancy suitable in the main for dry flies and also for those flies used with a sinking line. Buoyant flies and lures, when used with a sinker, tend because of their physical properties, to swim above the bottom, and any subsequent pull on the line forces the fly to dive. This type of fly can be very effective fished over weed beds, and the action of floating in mid-water and subsequently diving attracts the fish.

Plastazote Corixa (D. Collyer)
Hook 12-14
Body White plastazote which is glued to the hook shank and then cut to shape
Back & legs Cock pheasant tail fibre

Fishing the Corixa

The natural corixa, of whatever species, is active throughout the fishing calendar, so an artificial is worth a try right from the onset of the season. I have always found the silver corixa to be an excellent dropper fly when using a fish-imitating lure on the point. It can quite naturally be fished fast, as though being chased by our fishy lure; sometimes the fish are caught on the lure, at other times the trout are tempted to take the corixa pattern. I have no qualms about fishing this nymph-type fly at a fast rate. The natural moves like silver, greased lightning, so fish the artificial in the same way.

Other Water Bugs

ORDER: Hemiptera

THE WATER BOATMAN (*Notonecta glauca*)

After the corixa the next insect that could interest the trout, and therefore the angler/flydresser, is the Water Boatman. This insect is sometimes called the Backswimmer, because of its method of swimming, and sometimes it is referred to as the Wherryman. The Water Boatman is very much larger than its smaller cousins, and we have four species in this country, of which the most common is *Notonecta glauca*. It would be stretching the truth somewhat if I said that these insects were an important source of food for our friends the trout. But the Water Boatman is a creature of the smaller waters; I have seen many on some of the put and take fisheries, and I have found them in the stomach contents of caught fish.

Apart from the fact that they swim upside down they differ from the Lesser Water Boatmen in other ways; for one thing, they are highly predacious, unlike the Lesser Water Boatmen which are in the main herbivorous. The mouthparts of the *Notonectae* are developed into a beak-like form; this vicious mouth is stabbed into its prey, which can be anything from shrimps and larvae to fish much bigger than itself. They inject through this beak a toxic juice or saliva. This predigests its food, which is then sucked up, as though through a straw. So, beware! keep your fingers away from the mouth—the resultant bite-cum-stab is as painful as a wasp sting. You have been warned.

Description and Life Cycle

The Water Boatman is constructed in a similar manner to the corixa. The strong, highly fringed legs are a common factor in both species. The colour of *N. glauca* is a cinnamon brown on its elytra (wing cases), with a dark olive thorax, and it has the very large eyes typical of the species. Like the corixa it takes in air at the surface, but to do this it sticks its rear out of the water, unlike the corixa which pokes out its head. The air is trapped on the abdomen by two parallel rows of hydrofuge (unwettable) hairs; this bubble of air makes the insect extremely buoyant, and it is forced to cling onto anything that is available once below the surface. Although the corixa is found on the bottom when at rest, it is more usual to find the Water Boatmen hanging upside down at the surface, from which position it awaits its prey.

The Water Boatmen usually mate from December onwards, and eggs are laid around February. These eggs are deposited in slits in the stems of waterweeds. There are five nymphal stages, or instars, before the adult insect emerges. These instars, especially the fifth, are very green in colour (the earlier instars are whitish), and the creature itself far more rotund than the adult insect. The instar stages take approximately two months. The adult insect is a strong flier, and it flies during the hours of daylight. On one occasion a particular specimen thought the bonnet of my parked car to be some delectable pond shimmering in the noonday sun. I watched it whirl around and around on the hot metal, highly reminiscent of the Walt Disney film of the wild mallard landing on the frozen lake.

Artificials

This is the fly to be weighted along the back—that is, if you want to be dogmatic. As a counsel of perfection, tie the wing case beneath the hook rather than on top. But what a waste of time . . . the trout will not care which way up the fly swims.

Water Boatman (Price)
Hook 10
Silk Brown
Body Cream floss
Rib Wide flat tinsel
Back Cinnamon hen or turkey
Paddles As back, protruding out either side
Both the back and the paddles can be lacquered for durability.

There are very few, if any, published British patterns to choose from, so I have chosen two from the United States to recompense. Both look good fish takers. The first pattern can be used in the smaller sizes as a corixa pattern too.

The Water Bug (USA)
Hook 10
Silk Black or brown
Body Dark brown fur
Rib Copper wire
Back Mallard flank

Back Swimmer (USA)
Hook 14-10
Silk Olive
Body Olive tinsel chenille
Back & legs Dyed olive mottled turkey

These flies are to be found in the books, *American Nymph Fly Tying Manual* by Randall Kaufmann, and *Popular Fly Patterns* by Terry Hellekson. Both these books have a wealth of American fly patterns, many of which have proved their worth on British waters.

Randall Kaufmann, *American Nymph Fly Tying Manual*, Salmon Trout Steelheader (Portland, Oregon).

Terry Hellekson, *Popular Fly Patterns,* Peregrine Smith (Layton, Utah).

Fishing the Water Boatmen

Fish the artificial in much the same way as the corixa patterns. A case can be made, however, for fishing it more or less static in the surface film, for this is where the natural spends a lot of its time, hanging downwards surveying the aquatic scenery, awaiting the vibrational disturbance that signals food for this efficient killer of the ponds.

Before leaving the water bugs I must mention one that has been copied in the past by flydressers. The insect is the water bug *Velia capria* and the pattern tied to represent it called the Water Cricket. This insect favours slow flowing water, and really I should not have mentioned it in this book except for the fact that I happened to be reading my fishing diary and came across the following entry: *7th April 1969, Lake fishing/9 Browns, 1 Rainbow*. Amongst the four flies used that caught fish that day was the Water Cricket. Do you know, I have not used that fly pattern since. I must be mad. Probably I only had one, and lost it. I have now made a mental note to try it once more at the first opportunity next season. For those of you who would like to try it with me, the dressing is as given.

Water Cricket
Hook 12-14
Silk Orange or black
Body Orange floss silk
Hackle Starling back feather

Courtney Williams gives two patterns for the Water Cricket, but this one was the fly that worked the oracle for me. According to Courtney Williams the water cricket is always fished wet.